W9-CPE-614

Laboratory Manual

Biology of Microorganisms

Sixth Edition

Jane Ann Phillips/Thomas D. Brock

Laboratory Manual

Biology of Microorganisms

Sixth Edition

Thomas D. Brock
University of Wisconsin

Michael T. Madigan
Southern Illinois University

PRENTICE HALL
Englewood Cliffs, New Jersey 07632

Acquisition Editor: **David Brake**
Production Editor: **Maureen Lopez**
Prepress Buyer: **Paula Massenaro**
Manufacturing Buyer: **Lori Bulwin**
Supplements Editor: **Alison Munoz**

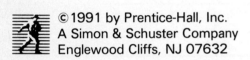
©1991 by Prentice-Hall, Inc.
A Simon & Schuster Company
Englewood Cliffs, NJ 07632

All rights reserved. No part of this book may be
reproduced, in any form or by any means,
without permission in writing from the publisher.

Printed in the United States of America

10 9 8 7 6 5 4 3 2 1

ISBN   0-13-083916-7

Prentice-Hall International (UK) Limited, *London*
Prentice-Hall of Australia Pty. Limited, *Sydney*
Prentice-Hall Canada Inc., *Toronto*
Prentice-Hall Hispanoamericana, S.A., *Mexico*
Prentice-Hall of India Private Limited, *New Delhi*
Prentice-Hall of Japan, Inc., *Tokyo*
Simon & Schuster Asia Pte. Ltd., *Signapore*
Editora Prentice-Hall do Brasil, Ltda., *Rio de Janeiro*

Contents

Preface vii
Laboratory safety ix

Exercises

1 Aseptic Transfer 1
2 The Microscope 11
3 Observation of Procaryotic Cells 23
4 Observation of Eucaryotic Cells 33
5 Growth 39
6 Environmental Parameters of Growth 49
7 Control of Microbial Growth 57
8 Biosynthesis and Nutrition: Catabolism 63
9 Biosynthesis and Nutrition: Anabolism 71
10 Culture Media 77
11 Identification of Bacteria 85
12 Bacterial Genetics 93
13 Regulation 103
14 Action of Bacteriophage 109
15 Microbes in Nature 115
16 Photosynthetic Bacteria 119
17 Soil Microbiology 125
18 Plant-associated Bacteria 137
19 Food Microbiology 143
20 Water Analysis 151
21 Microbes of the Body: The Cocci 159
22 Microbes of the Body: The Enterics 169
23 Identification of Unknown Bacteria 181

Student Supplements

Supplement 1 General Laboratory Procedures 187
   Labelling of cultures 187
   Record-keeping 188
   Sterilization of loops and other utensils 188
   Preparation of slides 188
      Wet mounts 189
      Hanging drop slides 189
      Slides for staining 190
      Heat-fixing a slide for staining 190
      Staining procedures 190
         The simple stain 190
         The Gram-stain—rapid method 191

The Gram-stain—conventional method 191

The negative stain 191

The spore stain 192

The acid-fast stain 192

The flagella stain 192

Enumeration of microorganisms by plate count 193

Pour plates 193

Spread plates 194

Streaking agar plates for isolated colonies 194

Streaking with a loop 194

Streaking with a cotton swab 196

Decontamination procedures 196

Laboratory tests 197

Amylase test 197

Catalase test 197

Coagulase test 198

β-Galactosidase test 198

Indole test 198

Methyl red test 198

Oxidase test 198

Serological tests for agglutination 199

Starch test 199

Tryptophanase test 199

Voges-Proskauer test 199

Supplement 2   Media and Reagents 201

Media 201

Reagents 211

Supplement 3   Dilution Theory and Problems 215

Appendices

Appendix 1   Media, Reagent, and Equipment Preparation 221

Media preparation 221

Preparation of glassware for sterilization 222

Sterilization using the autoclave 222

Sterilization by filtration 223

Reagent preparation 223

Appendix 2   Culture Preparation 225

Maintenance of stock cultures 227

Preparation of liquid cultures 228

Preparation of hay infusions 228

Preparation of Henrici slides 228

Preparation of *Clostridium* cultures 229

Preparation of concentrated water samples 229

Preparation of bacteriophage stocks 229

Preparation of unknown cultures 230

Appendix 3   Time Required for Each Exercise 231

Appendix 4   List of Vendors 233

Preface

This manual contains selected experiments for the introductory student. It has been written with the following criteria in mind:

- The experiments should be satisfactory for laboratory sections of 20–60 students.
- A variety of subjects and basic principles of laboratory microbiology should be presented.
- Maximal student safety should be ensured, with minimal use of potential human pathogens and harmful chemicals or procedures.

Although this is a completely new manual, all of the experiments presented here have been extensively tested with large numbers of students in our beginning microbiology laboratories at the University of Wisconsin—Madison. Instructors should be assured that all of the experiments, when done as indicated, work well. Where difficulties may be anticipated, we have so indicated clearly. It is highly recommended, however, that each exercise be pre-run by the instructor to identify any problems unique to a particular laboratory situation.

We believe this manual is unique in its practice of telling students what can go wrong, as well as what can go right, and we have used possible problems they may encounter to help illuminate basic principles. We have added questions throughout the manual as procedures are explained, in order for the student to be stimulated to think and to develop understanding. We have also included considerable detail on the preparation of culture media and on the purpose of various constituents of media, as we feel such in depth study will help the student understand various tests and procedures better. We have not written a *separate* instructor's guide to include culture and media preparation, but have included the necessary material as appendices to this manual. We feel it is in the students' best interest to have ready access to this information.

It is assumed that students using this laboratory manual will have access to a textbook of introductory microbiology, although we have not made references to specific textbook entries. We have, however, provided references to useful manuals of microbiological technique where more detail about particular techniques and procedures can be found. In addition, extensive cross-referencing has been made so that the order of presentation here need not be the order in which the experiments are actually done.

In our own laboratory teaching, it is our practice to begin each period with a brief discussion of the results and questions from the previous period, followed by an introduction to the current material. As much as possible, students should be encouraged to discuss the broader implications of the exercises in order to grasp more completely the concepts of microbiology.

Before beginning any laboratory work, the student should carefully read the introductory section concerning laboratory safety.

Finally, we wish to express our appreciation to Edward Phillips for the artwork, to Dr. Katherine Brock for her scientific and technical assistance, and to a long list of present and past laboratory instructors at the University of Wisconsin—Madison, including John Lindquist and Michael Westrick, for valuable discussions over the years.

Laboratory Safety

Safety in a microbiology laboratory is important in the prevention of infection that might be caused by the microorganisms being studied. In addition, many of the reagents and procedures used are potentially hazardous.

This manual does not require the use of any highly virulent human or plant pathogens. However, some of the organisms used are *potentially* pathogenic. This means that, although they may not cause disease in a normal healthy host, they might if the host were compromised. A human host can be compromised in a number of different ways: wounds and cuts, lowered resistance due to another disease, surgery, stress (including the stress of examinations), or immune-system disability (including auto-immune diseases or the use of immunosuppressive drugs). In addition, infection can occur, albeit rarely, by relatively nonpathogenic organisms even in healthy hosts.

In addition to organisms, there are some chemicals used in this laboratory which are potentially harmful. Finally, many procedures involve glassware, open flames, and sharp objects which can cause damage if used improperly.

Although none of the procedures or materials used in this laboratory are very dangerous, to avoid the problems that might rarely occur, the following precautions should be taken:

1. Follow precautionary statements given in each exercise. Situations which require extra care are identified by a caution marker in the margin.
2. Hands should be washed before and after any laboratory work. Hands should also be washed after cleaning up any spills (see below).
3. Long hair should be tied back so that it does not catch fire in the Bunsen burner flame and does not fall into sterile media.
4. Only absolutely essential materials should be brought into the laboratory. Coats, backpacks, purses and other material should not be placed at or on the laboratory bench. Alternative arrangements for storage of these items should be available.
5. Clothing worn in the microbiology laboratory should not be subsequently worn in a facility where there are compromised hosts, such as a hospital, clinic or nursing home, nor in an area of public food preparation. Preferably, laboratory aprons or coats should be worn in and not removed from the laboratory.
6. The bench-top should be wiped with a disinfectant both before beginning and after finishing laboratory work.
7. Equipment containing microbial cultures should be handled in a safe manner. Proper storage racks should be used, and cultures should be carried in small batches to reduce the risk of accident.
8. Spills should be cleaned up immediately. Spills of reagents should be cleaned using paper towels, followed by a complete rinse with water.

Spills of material containing viable organisms should be immediately contained with dry paper towel. The dry towel will soak up the spill, and so should be discarded in a proper container for sterilization. Following this, the area of the spill should be disinfected with bench disinfectant.

9. All cultures should be properly labelled and properly discarded after use. No slides, tubes, plates, flasks, or other containers with viable organisms should be washed or discarded until after they have been properly sterilized.

10. Spills, cuts and other accidents should be reported to the instructor in case further treatment is necessary.

11. Aerosols should be avoided by use of proper technique for flaming the inoculating loops and by performing any mixing of cultures and reagents in such a way as to avoid splashing (proper use of these procedures will be demonstrated).

12. Cultures or reagents should not be pipetted by mouth; a pipette bulb should be attached to the pipette.

13. Do not eat, smoke, drink or chew pens in the laboratory.

LABORATORY MANUAL

BIOLOGY
of
MICROORGANISMS

Aseptic Transfer

The practice of microbiology relies on the ability to work with **pure cultures**. A pure culture is one in which all the cells present are of one type. For example, a pure culture of *Escherichia coli* contains only *E. coli* cells. Similarly, a pure culture of a specific strain of *E. coli* contains only cells of that strain. As you will learn later, there is a certain amount of genetic variability in a large population of cells; however, the concept of a pure culture is still usable.

Aseptic transfer is required to prevent contamination of a pure culture when working with it. Since many manipulations are required to study the characteristics of a culture, it is imperative that these manipulations be done using methods known to minimize contamination.

For this exercise, you will be supplied with pure cultures grown in a rich medium called *nutrient broth*. The medium was sterilized using an autoclave before inoculation with the pure culture. A sterile medium is one in which all life is absent. The principles of medium preparation and sterilization are outlined in Appendix 1, which should be read carefully before beginning Exercise 1. During Exercise 1, be aware of the definition of the concepts of **sterility** and **pure culture**. You should also read the material in Student Supplement 1 pertaining to proper laboratory procedure.

Period A

The purpose of this exercise is to familiarize you with aseptic techniques. All the procedures have been designed to minimize contamination during transfer. You should follow these procedures exactly.

Materials

1. Overnight cultures of *E. coli, Staphylococcus epidermidis,* and *Bacillus cereus;* approximately 2 milliliters (ml) per student
2. An inoculating loop
3. Four tubes of sterile nutrient broth; approximately 5 ml per tube
4. One bottle of sterile nutrient agar, 100 ml per bottle; cooled at 50° C in a water bath
5. Four sterile glass or plastic petri plates
6. Clean microscope slides (1–3)
7. A wax pencil or other water-resistant marker

Procedure

1. Place the four petri plates right-side up in front of you. Obtain a bottle of melted and cooled agar from the water bath. Carefully wipe the adhering water from the bottle so that, when you tip the bottle, this contaminated water will not run into the plate. Light a Bunsen burner and properly adjust the flame. Remove the cap of the bottle as illustrated in Figure 1-1. Lightly pass the lip of the bottle through the flame to burn off any adhering dust. The lip of the bottle should already be sterile from the autoclaving procedure. Pour approximately one-fourth of the bottle's contents into each plate, being careful to prevent the drip on the lip from running down onto the contaminated outer glass, then returning to be poured into the plate. Carefully slide the plates to an undisturbed area of your bench to solidify.

2. Label one tube of sterile nutrient broth for each culture. Label the fourth tube "sterile control."

3. Holding your loop like a pencil, insert the loop into the flame as illustrated in Figure 1-2. The orientation of the loop wire in the flame is important for proper heating. Keep the wire in the flame until it is red-hot. Then move the adjacent nonwire part of the loop lightly through the flame. The wire will now be sterile, and the nonwire part will have any dust burned off that might have fallen into the media during the transfer procedure. Remember that the loop is now *hot and sterile*. Allow the loop to cool for a few seconds in the air.

4. Now pick up the tube of *E. coli* culture with your other hand, while still holding the *sterile* loop. With the hand holding the loop, remove the cover from the culture tube as shown in Figure 1-3. *Do not* put the cover down on your bench. If the tube was closed with a cotton or plastic plug, lightly pass the lip of the tube through the Bunsen burner to burn off adhering dust. If the tube was closed with a plastic cap, the lip of the tube should be sterile and this flaming is probably unnecessary.* Now, insert the loop into the broth, then remove it, carrying a loopful of culture. Replace the tube cover and return the tube to the rack.

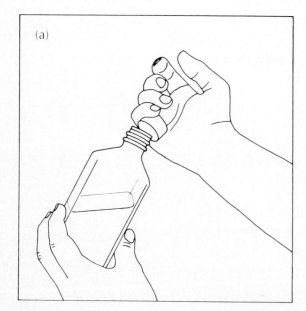

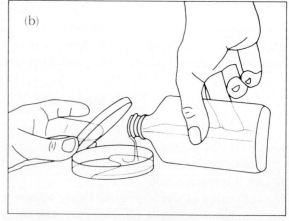

Figure 1-1 (a) *Removal of a bottle cap aseptically. Note how the rest of the hand is free to manipulate a pipette or plate lid; (b) pouring agar into a plate. Note how the lid shields the agar from airborne contamination.*

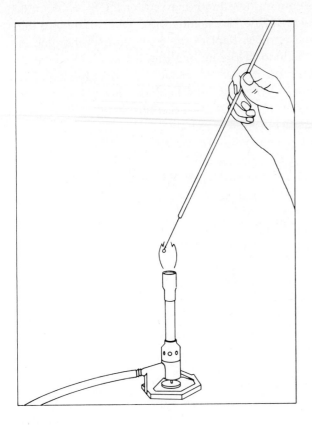

Figure 1-2 *Proper flaming of a loop. Note how the handle is gripped like a pencil, and the loop is inserted into the hottest part of the flame.*

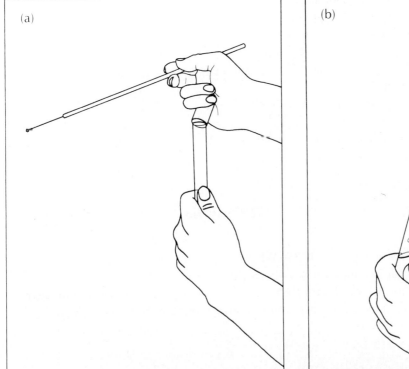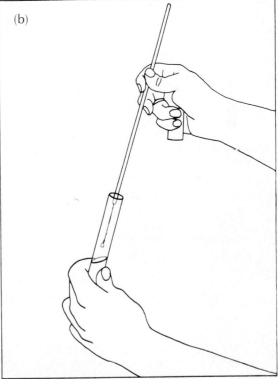

Figure 1-3 *Transferring a culture. (a) Removal of a tube cap while manipulating a loop; (b) obtaining inoculum from a broth tube while maintaining sterility of the cap (note cap in hand).*

5. Pick up the tube of sterile nutrient broth marked for the *E. coli* culture. Remove its cover, insert the loop containing the *E. coli* culture into the broth, swirl gently and remove. Replace the cover and set the tube in the rack. Resterilize the loop before putting it down to avoid contamination of the bench with the culture. *Be careful here!* When the loop has liquid in it, you must insert the loop into the flame *slowly* to allow the liquid to evaporate rather than boil, which would splatter live bacterial cells all over the bench, your books, and you.

6. Repeat steps 3, 4 and 5 with the other two cultures, transferring each culture into its appropriately marked tube.

7. Transfer a loopful of each culture to a clean glass microscope slide. With a glass-marking pencil, mark each slide on the bottom to indicate which smear is which. For example, draw a ring under the area containing the smear and write a code number or letter next to it. Be sure to keep a record of your code. Allow the slide to air-dry and save it until Exercise 2.

8. To test your ability to properly sterilize your loop, pick up a loopful of the culture of *B. cereus*, following the procedure of step 4. Properly flame sterilize the loop containing the *B. cereus* culture, then insert the loop into the broth marked "sterile control." Repeat this a few times with the same tube. If your flame sterilization technique is effective, the broth will be sterile next period.

9. Check to be sure the agar has solidified in your plates by carefully tilting the plates and looking for movement. If the agar is solid, label one plate on the bottom as "sterile control." Use the other plates to test for the natural contamination of articles in your environment. Suggested experiments include opening a plate to the air for a few minutes, coughing into a plate, rubbing a piece of your hair with your fingers and placing the hair into the plate, touching the agar surface both before and after you wash your hands, or rubbing the agar surface with any small article, like a coin. Be sure to label each plate to indicate the source of contamination.

10. Incubate both the tubes and the plates at 30° C until the next period. The plates should be incubated inverted to prevent any condensation from falling onto the agar surface, which would disrupt discrete colonies.

11. Discard all tubes and bottles as instructed.

Questions

1. What is a pure culture? What is its use in microbiology?
2. What is the definition of sterility?
3. Why is it important that the water on the outside of a bottle of agar be wiped off before beginning to pour the agar into a petri plate?

Materials

1. Plates and tubes incubated since last period
2. Inoculating loop and three sterile swabs for streaking plates
3. Microscope slides
4. Two nutrient agar plates, pre-dried for 2 days at room temperature

Procedure

1. Observe your tubes from the last period. Notice the turbidity in the tubes to which you transferred cultures. Record your observations on Report Sheet 1. Is there any turbidity in the tube marked "sterile control"? There should not be. If there is, ask your instructor to observe your transfer techniques to suggest what you are doing incorrectly. Prepare a slide of each culture as in Period A. (You should note that it is possible to put up to three cultures on a slide if you are careful.) Allow the slides to air-dry. Save them for Exercise 2.
2. Observe your plates from last period. Your sterile control should have no colonies on it. Your other plates will show colonies composed of large populations of cells as the result of growth and division of a single cell or clump of cells that landed on the agar surface at that spot. Record the diversity of colony size, shape, color, and number on the Report Sheet. Do not open these plates, as they could contain pathogens (better safe than sorry). Discard the plates as instructed.
3. Streak the nutrient agar plates using a three-phase streaking pattern (as shown in Figure 1-4) with the *E. coli* culture, using the inoculating loop for one plate and the sterile swab for the other.

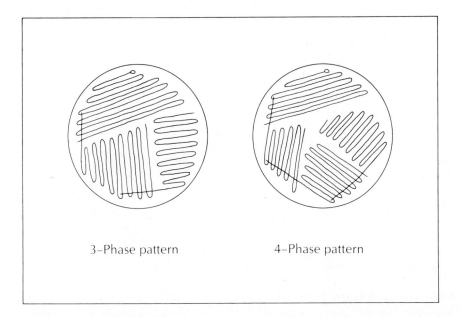

3–Phase pattern   4–Phase pattern

Figure 1-4 *Patterns used in streaking a plate to obtain isolated colonies. If an inoculating loop is used, it is sterilized between each phase.*

To streak a plate using the inoculating loop:

- Sterilize the loop and allow it to cool. Obtain cells from the *E. coli* culture, allowing the film of liquid inside the loop to "break" in the tube so that only those cells adhering to the wet wire and not those in a film within the wire are transferred.
- Put the cells at one edge of the plate as illustrated in Figure 1-4, then streak the culture on approximately one-third of the plate, leaving no open spaces. The handling of the plate can be accomplished in a number of ways, all of which attempt to minimize possible contamination by either keeping the lid over the plate or by keeping the plate upside-down (see Figure 1-5).
- Sterilize the loop and allow it to cool in the air for 15 seconds. Touch the loop to an unused edge of the agar surface to cool it completely before continuing. Then pull the loop through one edge of the previous streaks *one or two times* to re-inoculate the loop. Now streak the second third of the plate, *avoiding the first third* (see Figure 1-4).
- Sterilize and cool the loop as in the last step. Pull the loop through one edge of the streaks *in the second third of the plate* to obtain inoculum. Now streak the last third of the plate.

To streak a plate using the sterile swab:

- Remove the sterile cotton swab from its wrapping, being sure not to touch the swab to anything. Insert the swab into the culture of *E. coli*, then roll the swab on the side of the container to remove excess liquid. Using the swab like a loop, streak one-third of the agar surface as shown in Figure 1-4. Discard the swab as instructed.
- Obtain a second (and then a third) swab and proceed to streak the plate as would be done with a loop.

These directions for streaking a plate to obtain isolated colonies are also given in Student Supplement 1.

Incubate the plates in an inverted position at 30° C until the next lab period.

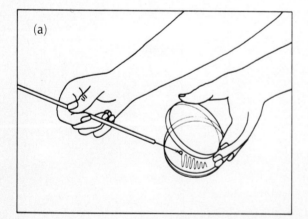
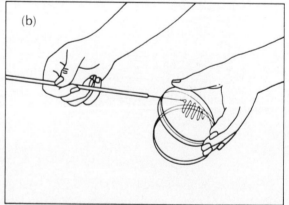

Figure 1-5 *Two procedures for streaking a plate. (a) Streaking a plate while holding the lid ajar. Note that the lid shields the agar from airborne contamination; (b) streaking a plate while holding the bottom of the plate. Note that the agar surface faces downward, thereby minimizing contamination from the air.*

Materials

Plates from Period B

Procedure

1. Observe the plates you streaked last time for isolated colonies. Show the plates to your instructor for comments. It is to your advantage to have the instructor look at your plates so you can learn the proper procedure as early as possible. If you have any problems, it would be useful at this time to obtain more plates and cultures to practice streaking.
2. Discard the plates as instructed. It should now be considered common practice to clean up properly after each lab period. Remember, this includes wiping your bench top with disinfectant.

References

Barkley, W. E. 1981. *Containment and Disinfection*. In **Manual of Methods for General Bacteriology**. Gerhardt, P., editor. American Society for Microbiology. Washington, D.C. A good reference not only for beginning students, but also for those performing more advanced laboratory work in microbiology.

*Some microbiologists flame all tubes as a matter of course. However, it is possible that the extra manipulation time with the cover off actually increases contamination rather than prevents it.

Notes

Aseptic Transfer

Culture transfer

Successful transfer

Turbidity in nutrient broth (+ or −)

Escherichia coli? _____

Staphylococcus epidermidis? _____

Bacillus cereus? _____

Aseptic technique

Any turbidity in control tube of nutrient broth?

Any contaminants on sterile control plate of nutrient agar?

If either of the above two answers are "yes", what are the possible points of contamination in the procedure?

(Continued on next page)

Distribution of microorganisms in nature

Observations of inoculated nutrient agar plates

Inoculum source	Morphology, size, and color of representative colonies
1.	
2.	
3.	

Streaking agar plates for isolated colonies

Did you obtain isolated colonies on the agar plates which were streaked with *E. coli*? If not, what changes should you make in your technique to ensure isolated colonies?

The Microscope

The microscope is one of the microbiologist's greatest tools. It allows for the visualization of small particles, including microbes which *individually* are too small to be seen with the human eye. With the help of proper illumination, a microscope can magnify a specimen and optically resolve fine detail. The important aspects of magnification, resolution and illumination are discussed below. This introduction to microscopy will include an explanation of features and adjustments of a compound light microscope. However, the technique of microscopy can be modified and refined so that it can have many more applications than the simple viewing of small particles. The student is encouraged to explore these other aspects (see references).

Before beginning to read the following discussion of the theory of the microscope, please familiarize yourself with the names of the microscope parts shown in Figure 2-1.

The Optical System The optical system of a compound microscope consists of two lens systems: one found in the objective; the other in the ocular (eyepiece). The objective lens system is found attached to a rotating nosepiece (see Figure 2-1). A microscope usually has three or four objectives that differ in their magnification and resolving power. **Magnification** is the apparent increase in size of an object. **Resolving power** is the term used to indicate the ability to distinguish two objects as separate. The most familiar example of resolving power is that of car headlights at night: at a long distance away, the headlights appear as one light; as the car approaches, the light becomes oblong, then barbell-shaped, and finally it becomes *resolved* into two separate lights. Both resolution and magnification are necessary in microscopy in order to give an apparently larger, finely detailed object to view.

Look at the engravings on the objective lenses and note both the magnification (for example: 10×, 45×, 100×) and the resolution (given as N.A. = numerical aperture, from which the limit of resolution, R, can be calculated: R = wavelength of light used divided by 2 × N.A.). The resolving power of the lens separates the details of the specimen, and the magnification increases the apparent size of these details so that they are visible to the human eye. Without both resolution and magnification, you would either see nothing (good resolution, no magnification) or a big blur (poor resolution, good magnification).

The objective lens system produces an image of the specimen, which is then further magnified by the ocular lens. The magnification of this lens is engraved on the ocular. The total magnification of the microscope is determined by the combination of objective and ocular in use. For example, with a 10× objective lens and a 10× ocular, the total magnification of the microscope is 100×. If the objective lens is changed to a 20× objective, then

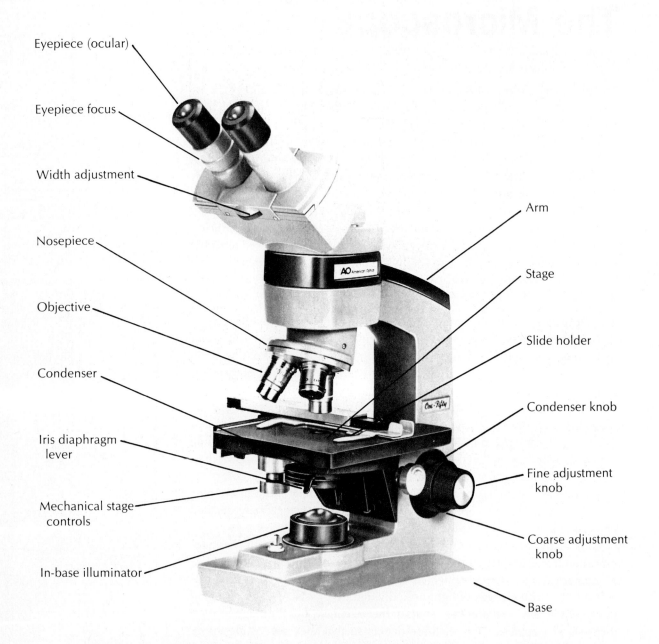

Eyepiece (ocular)

Eyepiece focus

Width adjustment

Nosepiece

Objective

Condenser

Iris diaphragm
lever

Mechanical stage
controls

In-base illuminator

Arm

Stage

Slide holder

Condenser knob

Fine adjustment
knob

Coarse adjustment
knob

Base

Figure 2-1 *The microscope. (Photograph provided by AO Scientific Instruments Division)*

the total magnification is now 200×, whereas if a 10× objective is used with a 12.5× ocular lens, the total magnification is now 125×. The use of objective and ocular lenses with different magnifications allows greater flexibility when using the compound microscope.

The Illumination System The objective and ocular lens systems can only perform well under optimal illumination conditions. To achieve these conditions, the light from the light source (bulb) must be centered on the specimen. (In most inexpensive microscopes, this centering is adjusted by the manufacturer. In more versatile microscopes, the centering becomes more critical and is a function performed by the operator.) The parallel light rays from the light source are focused on the specimen by the condenser lens system (see Figure 2-1). The condenser can move up and down to effect this focus. Finally, the amount of light entering the condenser lens system is adjusted using the condenser diaphragm. It is critical that the amount of light be appropriate for the size of the objective lens receiving the light. This is important to give sufficient light, while minimizing glare from stray light which could otherwise reduce image detail. The higher the magnification and resolving power of the lens, the more light is needed to view the specimen.

Objective lenses used for observing very small objects such as bacteria are almost always *oil immersion* lenses. With an oil immersion lens, a drop of oil is placed between the specimen and the objective lens so that the image light passes through the oil. Without the oil, light passing through the glass microscope slide and specimen would be refracted (bent) when it entered the air between the slide and the objective lens. This refracted light might still be able to contribute to the image of the specimen if the objective lens is large. However, at the higher magnification, the objective lens is small, so is unable to capture this light. The loss of this light leads to loss of image detail. Therefore, at higher magnifications, the area between the slide and the lens is modified to have the same (or nearly the same) refracting qualities (**refractive index**) as the glass and specimen.

To use an oil immersion lens, place a drop of oil on top of the coverslip (or dried specimen) and *carefully* focus the microscope so that the objective lens is immersed in the oil. Any lens which requires oil is marked "oil" or "oil immersion." Conversely, any lens *not* marked "oil" should *not* be used with oil and is generally not sealed against oil seeping into and ruining the objective.

General Operating Procedure

The steps taken when using a compound microscope are outlined here. Go through these steps now, and again when carrying out the procedures for Period A.

1. Plug in and turn on the microscope light source. If there is a variable voltage transformer, set it at 2–3 volts for low magnification and increase up to a maximum of 6 volts as needed for higher magnification.
2. Looking from the side, raise the objective nosepiece by means of the coarse adjustment knob.
3. Rotate the nosepiece and set in place the objective with the lowest magnification. You should feel a definite position stop for the objective.

Turn the nosepiece clockwise and counter-clockwise until you are familiar with this position stop.

4. Place the slide in the stage slide holder securely. (Be sure the slide has the *specimen side up*.) Roughly center the specimen over the light coming from the condenser.

5. Lower the objective using the coarse control knob until it reaches a stop. *Do not* force the knob. The stop should be obvious if you are moving the controls gently. Again, become familiar with the *feel* of the position stop.

6. Position the condenser about a paper-thickness below its upper-most position.

7. Looking through the ocular(s), focus the specimen image using the *fine* control knob. (Your first time through, you might find a red wax pencil line makes an easily viewable specimen.)

   If you have a binocular microscope, focus using *one eye only* (usually the right); then close the right eye and focus the image seen by the left eye using the *ocular adjustment* (not the focus knobs). This adjustment corrects for focus differences between your two eyes. Finally, looking with both eyes, use the knurled knob usually found between the two oculars to adjust the distance between the oculars to match your interpupillary distance.

   Most microscopes used in classrooms have *high eyepoint* oculars which allow people who wear eyeglasses to keep them on.* Thus, if you are not wearing glasses, the focus point for comfortable binocular viewing will be further back than you might expect (your eyelashes should not be hitting the ocular). Still having trouble seeing one image using both eyes? Try looking out the window or across the room for a minute or two to relax your eyes. Many first-time users expect to have to focus as if the image were close to their eyes. However, the optics of the microscope are designed to that the apparent distance from the image to the eye is about the same as that to a book you might read comfortably for hours. This allows you to look from microscope to notebook and back without eye-straining focus changes.

8. Now remove an ocular. Put it upright in a *safe* place (away from elbows, books). Looking through the ocular tube, open or close the substage (condenser) diaphragm until the edges of it just leave the field of view. This is the proper adjustment of the light for best resolution. Replace the ocular.

9. Try the next magnification (40–45×). If your microscope objectives are **parfocal**, you need only rotate the nosepiece to position the objective. The specimen still should be in focus, or close to focus. If your objectives are *not* parfocal, repeat steps 2, 3, 5, and 7. Step 8 should always be repeated for *each* objective.

10. Try the highest magnification (100×), which is an oil immersion lens. If your objectives are parfocal, turn the nosepiece so no objective is in place, add a drop of oil, then rotate the oil immersion lens *into* the oil. The lens should not be touching the slide itself. If your objectives are *not* parfocal: raise the nosepiece using the *coarse* adjustment knob, add the oil, rotate the oil immersion lens into place and *looking from the side*, carefully lower the lens into the oil. Viewing through the ocular now, use the *fine* adjustment knob to focus the specimen image. This focusing should be done gently. Most modern microscopes have spring-loaded

objectives and/or nosepieces to prevent damage to the lens if the objective touches the slide. However, it is still necessary to be careful when focusing an image with a high-power objective, so be gentle and aware of the feel of the objective against the slide.

11. Ready to change slides? If you rotate the objective out of position, place the new slide in the stage clips, and return the objective back to position (adding oil first, if you are using the 100× lens), you should be in approximate focus, and you need only use the fine adjustment knob to sharpen the image.

12. Work completed? Clean all external objective and ocular lenses with **lens paper**, taking special care to remove all oil from oil-immersion lenses. *Do not* use solvents such as xylol! These solvents may dissolve the adhesive holding the lenses in place. Clean the stage, if necessary. Wrap the electrical cord as instructed. Return the microscope to its resting place, being sure to always carry the microscope upright with two hands: one hand holding the arm of the microscope, one supporting the base.

Problems? Various problems might arise as you work with your microscope. If you cannot bring the specimen into proper focus:

- Be sure your slide is *specimen-side up*.
- Recheck your coarse-control knob. For simple microscope work with regular slides, the coarse-control knob should be turned so that the nosepiece is in the lowest position.
- Clean your objective lens with lens paper moistened with distilled water. Dry the lens with more lens paper.
- If the specimen is in focus, but appears indistinct or difficult to view, clean the objective. Dirt on the objective seriously degrades the quality of the image. Clean your objective lens with lens paper moistened with distilled water. Dry the lens with more lens paper.
- If material appears in the image which does not seem to belong with the specimen, examine the ocular for presence of dirt. Dirt on the ocular lenses generally appears in the image as well-focused particles. To determine if the image is affected by dirt from the ocular, slowly rotate the ocular while looking through the lens. Dirt on the ocular will appear to move as you rotate the lens. Clean the ocular lens as you would an objective lens.
- If you are using an oil immersion lens, is there enough oil to span the distance between the slide and the objective lens?

If you are having difficulty getting enough light:

- Check the condenser. For lower magnifications, it should be about a paper-thickness lower than its highest point under the stage. For higher magnifications, you might need to raise the condenser slightly to ensure that the light is focused on the specimen.
- Check the condenser diaphragm. Be sure it is properly adjusted (opened wider for higher magnifications, closed down for lower magnifications).
- If you have a variable voltage transformer, check to see if you have adjusted the voltage upwards to increase the light intensity as the magnification increases.
- If you are using an oil immersion lens, is there enough oil to span the distance between the slide and objective lens?

Any other problems? Burned-out light bulb? Focus knobs grinding as they move? Focus knobs moving by themselves? *Do not attempt to fix these yourself! Get your instructor!*

Period A

In this period, you will learn to perform simple stains and to become familiar with the microscope.

Materials

1. A microscope
2. The slides prepared in Exercise 1
3. A dropper bottle of crystal violet or methylene blue stain to use for staining the bacterial cells

Simple staining procedure

In the first part of this period, simple staining procedures will be carried out. The purpose of staining is to increase the contrast between the organisms and the background so that they are more readily seen in the light microscope.

1. Light your Bunsen burner and adjust the flame. (After this, it will be assumed you know when your burner is needed.)
2. Holding a slide (specimen-side up) at one end with either your fingers or forceps, slowly pass the slide through the flame to warm the slide *slightly*. This process is called **heat-fixing** the specimen to the slide. Its purpose is to bind the specimen to the slide so that it does not wash off during staining. The slide should be warm to the touch, *not hot*. If you think you have heated the slide too much, *do not* touch the slide to your hand to find out. Chances are, your suspicions are correct, and you will burn your hand with the hot glass. Instead, heat-fix the slide for a shorter period of time next time, then test. Alternatively, you could slowly lower the slide *toward* your hand until you can feel the heat. A properly heat-fixed slide will not radiate enough heat to be felt until touching or almost touching the hand.

Safety Caution!

3. Stain the smear(s) on the slide by flooding the slide (again, specimen-side up) with either of the two stains listed above. This should be done over a sink, if possible, or at least over a beaker. Let the stain sit for one minute. Using a pair of forceps or a clothes pin, hold the end of the slide and tilt it to allow the stain to drain off. Now, rinse the remaining dye off with a gentle stream of water from a faucet or wash bottle. Let the slide air dry or blot it dry with blotters or paper towel. Blotters are preferred to paper towel as the towel often leaves paper fibers behind which might interfere with viewing. Be sure to *blot* the slide, *not* rub it.

Microscopic observations of stained specimens

Observe the slides prepared in Exercise 1 which have just been stained and record your observations on Report Sheet 2. Important observations include the shape and relative size of the three different cultures. *E. coli* cells are small and rod-shaped. You will be unable to see any detail within the cells. *S. epidermidis* cells should appear to be round (cocci) and appear in clumps. The cells should be similar in size to *E. coli* and should also be devoid of detail. *B. cereus* cells will be larger than either of the other two and will be rod-shaped. Depending on the age of the culture, you might be able to see an internal, nonstaining object in some of the cells. These are endospores and will be studied in more detail in Exercise 17.

If your aseptic transfer was successful, you should observe the same cells in your smear from the transferred culture as in the original. It is more likely that the *B. cereus* endospores will be visible in the transferred culture if the interval between transfer and smear preparation is more than a day apart. If you do not see the same cells in both smears, what happened? Can you identify the source of contamination? Is it one of the other cultures or possibly something from the air or tube? If your culture(s) was contaminated, be especially careful the next time you transfer cultures. If you are having difficulty, it would be useful to you to have your instructor observe your technique and make suggestions for improvement.

Questions

1. What is the difference between magnification and resolution? Why are both necessary for microscopy?
2. What is the purpose of the condenser? What is the result of having it adjusted improperly?
3. What is the purpose of the condenser (substage) diaphragm?
4. What is the function of the oil used with the oil immersion lens? What is the effect of omitting it? Why is it imperative to promptly remove any oil that has accidentally gotten on the other, non-immersion lenses?
5. Describe the lens systems of a compound microscope.
6. A simple microscope has only one lens system, whereas the compound microscope has both the objective and ocular lens systems. What is the advantage of having two lens systems?

Period B

In the previous period, you used the microscope for the first time, viewing organisms which had been stained so that the contrast between them and the background was increased. However, these stained organisms were dead. Now that you are familiar with the microscope, you will attempt to view living, unstained microorganisms. In order to see unstained organisms with the microscope, the light must be carefully adjusted. To do this, set the condenser diaphragm as before, then *close it down slightly*. The amount to close down the diaphragm will be obvious as you view the specimen. This adjustment of the diaphragm results in higher contrast between the microbes and their aqueous suspending liquid. As you view these living specimens, take note of the variety of sizes, shapes, and arrangements and look for the presence of motility.

In order to observe living microorganisms, you will prepare *wet mounts* and *hanging drop slides*. When making wet mounts, it is essential that the loop be cool before it is put into the cultures. Otherwise it will sizzle the cells. As you work with the two different methods of observing live organisms in liquid culture, think about the possible advantages and disadvantages of each. Then refer to Student Supplement 1 for a discussion of the two methods.

Materials

1. One hay infusion per four students
2. An overnight culture of each of the following organisms: *E. coli, Saccharomyces* sp. and *Bacillus megaterium*†
3. One hanging drop slide (depression slide)
4. Coverslips

Procedure

1. Remove a loopful of *Saccharomyces* culture and place it on a clean glass slide. *Do not* let it dry, but rather place a clean coverslip on top of the liquid. Using blotters or paper towels, blot the excess liquid from the slide.‡

2. Place the slide in the microscope slide clips *coverslip-side up*. Center the specimen over the circle of light on the stage. If you have air bubbles, align the slide so the edge of an air bubble seems to bisect the cone of light from the condenser. Close down the condenser diaphragm slightly. Focus on the edge of the air bubble, or on the *Saccharomyces* cells. If you have focused on the air bubble, look on both sides to determine which side is the liquid that contains the cells. Adjust the condenser diaphragm for maximum contrast.

   Proceed to 45× magnification. Readjust the diaphragm, if necessary. Note the size of these cells and the difference between them and the procaryotes you observed in Period A. Note the presence of internal organization. The largest structure inside the cell is a membrane-enclosed vacuole. *Saccharomyces* is a budding yeast, that is, one which divides by budding off from the parent cell. Look for the presence of *buds* at the ends of some cells. Record your observations.

   Proceed to 100× magnification. The drop of oil should be placed directly on the coverslip. Readjust the condenser and condenser diaphragm to give maximum contrast. Can you see any more intracellular detail? Record your observations.

3. Prepare a wet mount of *B. megaterium*, a relatively large procaryotic cell. Compare what you see to your observations of *Saccharomyces*, the eucaryote. Record your observations on Report Sheet 2.

18

4. Prepare a wet mount of *E. coli*. Note the size difference between it and the eucaryotic yeast. Eucaryotes are almost always larger than procaryotes. Looking at *E. coli* again, note the lack of obvious internal organization. Of course, this apparent lack of intracellular organization is actually due to the inability of the microscope to resolve it in such a small cell. Record your observations.

5. To view the material in the hay infusion, you may use a wet mount. However, when there is a considerable amount of solid matter present (such as hay particles), it is difficult to make a good wet mount since the coverslip just lies on top of the solids. Therefore, another technique, the hanging drop slide, is often used. To prepare a hanging drop slide, place a small dropful of material from the top of the hay infusion *on a coverslip* placed on the bench. Then pick up the hanging drop slide. Place a small drop of water, oil or Vaseline on the flat area on the edge of the depression (see Figure 2-2). Now place the slide *depression side down* on top of the coverslip, centering the drop on the coverslip over the center of the depression. Quickly and carefully, invert the slide. The coverslip should stick to the slide, and the loopful of material from the coverslip should now be *hanging* in the depression. Put the slide into the microscope slide clips coverslip-side *up*. The hanging drop slide is thicker than a normal slide, so you should be especially careful when focusing to prevent damage to the objective lenses. Focus on the edge of the drop using the $10\times$ objective. You will note that when you are in correct focus, the coarse adjustment knob is *not* all the way to its stop position. Maintain your focus even with the edge of the drop, as attempting to focus on the bottom of the drop with the higher magnifications will require that the objective be so low that it might break the coverslip and possibly damage the outer objective lens.

Now, make and record observations of the life in a hay infusion. Adjust the condenser diaphragm as needed for optimal contrast. Can you identify possible procaryote and eucaryote cells by their sizes and internal structures? Can you see any greenish cells (algae)? Watch the protozoa. Can you see the "mouth" (oral groove) used to ingest prey, the bacteria? Watch the internal structures of the protozoa. Do they appear to be changing as the cells ingest and process their food? At the higher magnifications, can you see any cellular arrangements for the procaryotes

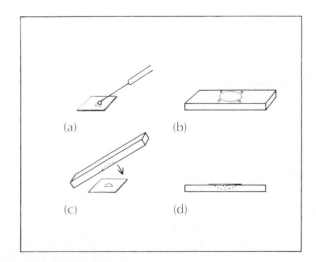

(a)   (b)   (c)   (d)

Figure 2-2 *Preparation of a hanging-drop slide. (a) A drop of culture is put onto a coverslip; (b) the hanging-drop (depression) slide is prepared by putting small drops of water or Vaseline at four corners around the depression; (c) the hanging-drop slide is inverted over the coverslip; (d) the completed hanging-drop slide is turned upright. Note how the drop of culture now hangs in the depression.*

(clumps, pairs, chains)? Do any of the cells seem to be associated with hay particles? With algae? Why do you think the bacteria would strongly associate with these materials?

After you have looked at one hay infusion slide, try preparing a slide from material near the bottom of the hay infusion. Do you see any differences in the microflora? If so, speculate as to the possible reasons. Later in the semester, you may wish to return to your observations here to see if your initial speculations were reasonable.

The viewing of life in a hay infusion can be very exciting and very instructive. Imagine the thrill the early scientists had when they first observed these "animalcules" in lake and rain water. As Antoni van Leeuwenhoek wrote in 1680: "All we have yet discovered is but a trifle, in comparison of what still lies hid in the great treasury of Nature."

6. Dispose of the wet mount slides as directed. The hanging drop slide is reusable, so simply discard the coverslip, clean the slide and return it to its proper storage place. If the hanging drop slide has been used with material likely to contain potential pathogens, it should be sterilized in disinfectant before cleaning. A hay infusion made as directed is not considered a likely source of pathogens.

References

Hartley, W. G. 1964. **How to Use a Microscope**. The Natural History Press. Garden City, New York. Unfortunately, this book is out of print, but is available in some libraries. It is a complete and understandable book of both the theory and practice of all phases of light microscopy.

Murray, R. G. E. and C. F. Robinow. 1981. *Light Microscopy*. In **Manual of Methods for General Bacteriology**. Gerhardt, P., editor. American Society for Microbiology. Washington, D.C. A short, easy-to-read treatise on the most common uses of light microscopy. Includes some troubleshooting.

*A word about glasses. Since a microscope can be focused, it is capable of correcting for most near- or far-sightedness. Thus, those with corrections for near- or far-sightedness alone can choose not to wear their glasses. Microscope lenses are *not* capable of correcting for astigmatism, so those having corrections for astigmatism should wear their glasses. The eyeglasses need *not* touch the oculars for proper viewing, so individuals with plastic lenses should not worry about scratches.

†One to two ml per four students should be sufficient.

‡Excess liquid causes the coverslip to float. When using the oil immersion lens, the coverslip will stick to the oil rather than the slide, making viewing nearly impossible. The excess liquid also results in viewing difficulty as motile organisms can easily move up and down in the liquid, moving into and out of the field of focus. "If I have air bubbles, should I re-do the mount?" No. Air bubbles are actually very useful for initial focusing.

The Microscope

Simple stain of cultures from Exercise 1

Morphology and relative size of cells

Culture	Observations of smear from original culture	Observations of smear from transferred culture
Escherichia coli		
Staphylococcus epidermidis		
Bacillus cereus		

If you transferred each culture aseptically, the results from each smear should be the same for any particular culture (although the *B. cereus* culture may have more spores in the transferred culture). Was your aseptic transfer in Exercise 1 successful?

Wet mounts

Culture	Cell morphology
Saccharomyces sp.	
Bacillus megaterium	
Escherichia coli	

(Continued on next page)

Observations of the hay infusion

Sample from the top of the hay infusion:

Sample from the bottom of the hay infusion:

Observation of Procaryotic Cells

In the last exercise, you looked at a few different kinds of procaryotic cells. You were asked to pay attention to the cell shape, size and arrangement. These three parameters are very useful when attempting to identify an unknown organism or when checking a culture of a known organism for purity. Some examples of cell shape and arrangement are given in Figure 3-1. In addition, the use of **differential staining** can be helpful to further identify characteristics of the cells. The most widely used differential stain is the **Gram stain**. This staining procedure, described in detail below and in Student Supplement 1, results in two different colors of stained cells: purple cells, which are termed **Gram-positive** because of their ability to retain the crystal violet dye after **decolorization** with alcohol; and red or pink cells, which are termed **Gram-negative** because of their inability to retain the purple dye. The Gram-negative cells are visualized by using a red **counterstain** called safranin. The Gram stain has been used as a taxonomic tool for

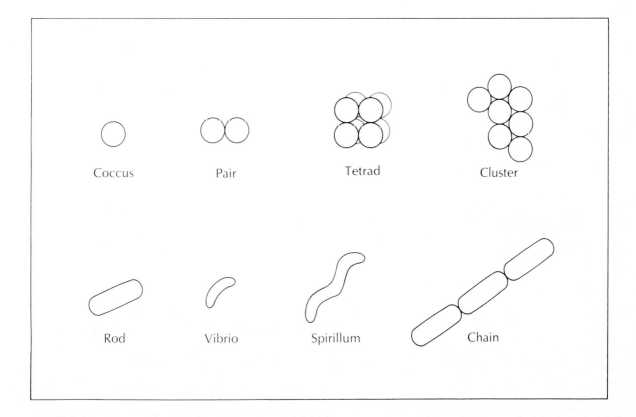

Coccus	Pair	Tetrad	Cluster
Rod	Vibrio	Spirillum	Chain

Figure 3-1 *Morphology and cellular arrangement seen in commonly isolated bacteria.*

many years. It has only been recently (relatively speaking) that the true basis of the differential reaction to the procedure has been determined: the cell walls of the two groups are morphologically and chemically quite different. For a complete discussion of the cell wall differences, you are referred to your textbook. Some factors do influence the staining of Gram-positive cells: the age and the pH of the culture. Older cells often lose their ability to retain the crystal violet dye, and the population of cells thus becomes *Gram-variable* (both pink and purple cells present) or Gram-negative. Young cultures must therefore be used for the Gram stain procedure. Cultures are also grown for the Gram stain in media low in sugars to avoid the formation of acidic endproducts during cell growth.

Another type of differential stain is the **negative stain**, which allows for the visualization of any *large* capsules around the cell. This stain uses India ink, which is a suspension of black carbon particles. In a wet mount, the particles will surround but not penetrate the cells. Cells which have large capsules around them will be visible inside a white "halo" (see Figure 3-2). This halo is due to the capsular material which prevents close association of the India ink particles with the cell wall. Cells without large capsules lack this halo. Since many organisms only produce large capsules under certain environmental conditions, the lack of a capsule is not proof that the cells cannot produce a capsule. However, using *standard* growth conditions, the presence or absence of a capsule can be a useful taxonomic parameter.

The **spore stain** is yet another example of a differential stain. The spore stain is used to stain bacterial endospores produced by members of the genera *Bacillus* and *Clostridium*. These structures are impervious to most ordinary stains. Therefore, to stain them, a dye such as malachite green or carbol fuchsin is placed on the smear and heat is used to drive dye into the spores. The excess dye is rinsed off, and the unstained vegetative cells are *counter-stained* with an appropriate dye of a different color, such as safranin or methylene blue. An example of what might be seen using malachite green and safranin for the staining procedure is shown in Figure 3-3.

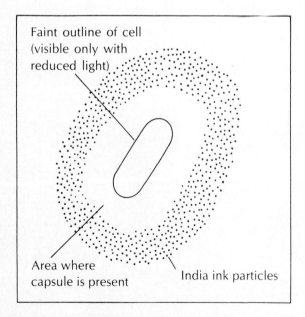

Figure 3-2 *Results of a capsule stain.*

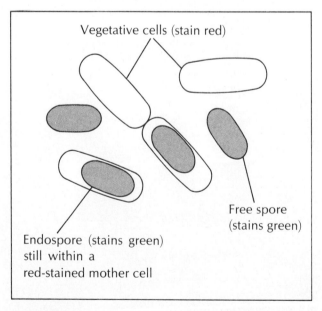

Figure 3-3 *Results of spore-staining using malachite green with a safranin counterstain.*

Some organisms have distinctive cell walls which can be identified for diagnostic purposes by using an appropriate differential stain. Probably the best known example of this is the **acid-fast stain** used to differentiate *Mycobacterium* cells from cells of other bacteria. The acid-fast stain is similar to the spore stain in that heat is often used to impregnate the cells with the dye. Cells which retain the dye after decolorization with acid-alcohol are said to be **acid-fast**. Since *Mycobacterium tuberculosis* and *M. leprae* are causative agents of human disease, tuberculosis and leprosy, respectively, it is useful for diagnostic purposes to have a rapid test for these organisms in clinical samples from diseased individuals. Some *Mycobacterium* species are normal inhabitants of the human body, so caution is required in diagnosing a disease based *only* on the presence of these acid-fast organisms.

Finally, another differential stain which will be demonstrated is the **flagella stain**. This stain coats the thin bacterial flagella with heavy metals or other compounds to make them visible in the light microscope. Once visible, the location and number of the flagella can be used diagnostically (see Figure 3-4). The presence of flagella varies with cultural conditions, so (as with capsules) a negative result is not proof of a cell's inability to produce flagella.

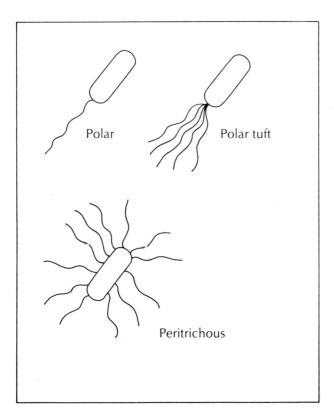

Figure 3-4 *Distribution of flagella on a cell.*

Period A

Materials

1. Overnight cultures of *E. coli, Proteus sp., S. epidermidis, B. cereus* and *Spirillum sp.* Cultures should be grown in shake flasks at 30° in media low in sugars.
2. Demonstration slides of the spore, acid-fast and flagella stains

Procedure

1. Prepare smears of the five bacterial cultures listed in #1 above. Be sure to mark the area and the identity of each smear carefully on the *bottom* of the slide. Allow smears to air-dry as you perform the next step.
2. Prepare a wet mount of the *Proteus* culture. Observe the preparation for motility *immediately* after preparation since the cells might lose their motility as oxygen is depleted under the coverslip. The cells should be moving with obvious *individual* direction; that is, if you follow any individual cell, it can move in the liquid in directions other than those followed by the cells around it. If the culture is very healthy, nearly all the cells will show motility. More likely, less than 100% of the cells will actually be motile. Now prepare a wet mount of the *S. epidermidis* culture. This organism is *nonmotile*, so any movement you see is due to passive flow with water currents (especially around air bubbles) or *Brownian motion*. Brownian motion is a vibrating-type movement caused by the dynamic motion of molecules in the suspending liquid. (It is analogous to the type of movement you would see from above when viewing beachballs in a swimming pool. None of the balls are capable of directing their own movement, but they bob and move in the water.) Compare what you see with your observations of the motile *Proteus* culture. Record your observations on Report Sheet 3. Prepare wet mounts of and observe the other cultures provided.
3. As time permits, view the demonstration slides of the flagella stain.
4. As time permits, also view the demonstration slides of the spore and acid-fast stains. You will be performing the spore stain later in the course.
5. After the slides prepared at the beginning of the period are dry, heat-fix them, then put them into your slide box until next period.

Period B

In this period, you will be performing the negative stain and the Gram stain. The Gram stain outlined below is the rapid method. It has gained widespread use in the last few years because of its ease of use and reproducible results. During the course, you might also try the longer method given in Student Supplement 1.

Materials

1. Slides prepared last period
2. Gram reagents, India ink designed for free-flow graphic pens
3. A 36–48 hour old culture of *Klebsiella pneumoniae* inoculated as a single streak across the middle of the plate of Eosin-Methylene Blue agar and incubated at 37° for 36–48 hours
4. Demonstration slides of the spore, acid-fast and flagella stain, if students have not finished viewing them

Procedure

1. To perform the Gram stain, place the slide over a beaker or sink. (Be sure the slides were heat-fixed last time.)

   - Flood the smear with Gram's crystal violet. (Avoid getting this dye on your clothes, books and fingers, as it is very difficult to remove. Use all dyes with caution.).
   - Add 3–8 drops of 5% sodium bicarbonate solution. Allow to act 5–10 seconds.
   - Rinse the slide with Gram's iodine, then flood the smear with the iodine. (This *mordants* or fixes the dye in the cells.) Allow it to act for 5–10 seconds.
   - Rinse the slide with alcohol-acetone, then flood it with the alcohol-acetone. Allow to act 5–10 seconds. This removes the crystal violet from Gram-negative cells. Pour off the excess reagent and allow the slide to dry for 15 seconds. The slide need not be dry to continue.
   - Flood the slide with safranin and allow to act for 10 seconds. This dye is a counter-stain which stains the Gram-negative cells pink.
   - Rinse the slide with water and *blot* dry. As you become proficient with this staining procedure, you will notice that the times required for action by the reagents are approximately the times required for you to manipulate the dropper bottles of reagents.

2. When the stained slides are completely dry, observe them using the highest magnification. If you find it difficult to get the smear in focus, use the wax-pencil focusing trick, where you perform your focusing on a wax-pencil line (which is, of course, put on the same side of the slide as your smear). After getting the wax-pencil line in focus, move your slide over to the area of your smear. You should be in approximate focus. Look for color as an indication of the presence of the smear, then sharply focus the cells using the fine adjustment knob. As you become more proficient at viewing stained cells, you will find you need the wax-pencil "trick" less and less. Observe the smear for cell shape, arrangement and Gram reaction. Record your observations on Report Sheet 3. Notice that the report does *not* call for cell color, but rather for Gram reaction. This requires an interpretation of (and confidence in) your actual visual observations. In another laboratory period, you may be given an unknown organism to stain, so you might wish to check your interpretations with your instructor today.

3. Prepare a negative stain of *Klebsiella*. To do this, suspend a small amount of cell material from the plate into a loopful of water on a clean glass slide. Add a loopful of India ink to one edge of the drop, allowing the drops to touch and mix. Now add a coverslip and blot as you would a wet mount. You should see a mount that varies in darkness from very dark on the edge where the India ink was applied, to much lighter on the other edge. Place a drop of oil on the coverslip and view the center of the slide in the microscope using the 100× lens. In the background, you should see the India ink particles. The cells themselves will appear to be slightly grey and surrounded by a white halo. If you can see the India ink particles and halo but not the cells, try closing the microscope iris diaphragm a small amount. A drawing of capsulated cells is shown in Figure 3-2. If all you see is black, move the slide to the lighter area. If all you see is white, move the slide to the darker area. If you are unable to see capsulated cells, prepare another slide, allowing more mixing of the two drops.

4. If you have not previously viewed the demonstration slides of the spore, acid-fast and flagella stains, do so.

Questions

1. The Gram reaction is used as taxonomic tool in the identification of bacteria. All the organisms in a given genus should have the same Gram reaction. Based on this knowledge, what is the Gram reaction of *Staphylococcus aureus*? *Bacillus subtilis*?

2. Capsules often interfere with the Gram reaction of a cell. Which of the organisms used in this exercise would you expect trouble with? The capsule of this organism is usually formed only under cultural conditions of high sugar. Looking in Student Supplement 2, what compounds would you omit from the medium on which this organism was grown to avoid capsule formation?

3. Can you distinguish between *E. coli* and *Proteus* by shape and Gram reaction alone?

Period C

In this period, you will be given an unknown culture to identify for Gram reaction, shape, cell arrangement and motility. You will also be given two cultures as "controls" for the various reactions. The results of your final determination should be turned in at the end of the next period.

Materials

One to two ml of *E. coli*, *S. epidermidis*, and an unknown

Procedure

1. Prepare a slide containing *all three* cultures. The two known cultures will serve as negative and positive controls of your Gram-staining technique. Let the slides air-dry.
2. Prepare a wet mount of your unknown culture and observe it for motility.
3. Make a Gram stain of the slide prepared in step 1. Observe the cells for shape, Gram reaction and arrangement. Record your results on Report Sheet 3.

Questions

1. If your unknown is a nonmotile, Gram-positive coccus arranged in clumps, can you *positively* identify it as *Staphylococcus epidermidis*? (If this helps any, could you positively identify a large, woody-trunked, green-leaved plant as an oak? What does this tell you about the number of characteristics that are necessary for positive identification of organisms?)
2. Now that you have completed this exercise, go back and review the uses of a stain, a decolorizing agent, a counter-stain, and a differential staining technique.
3. What would be the result of forgetting the counter-stain in the Gram stain technique?
4. What would be the result of decolorizing too long in any staining procedure which uses a decolorizing step?
5. You have been given a sputum sample from a patient who has been having respiratory problems. You prepare a slide and stain it with the acid-fast stain. You see acid-fast rods among many other cells which are not acid-fast. Can you conclude that the patient has tuberculosis? Why or why not?
6. Why is heat used in the staining procedure for the spore and acid-fast stains?
7. In the spore stain, you observe a green oval structure within a larger pink structure. What is the green structure? What is the pink structure?
8. In the spore stain, you observe a green oval structure completely separate from any pink structure. What is it?
9. You and your friend decide to split up some of the laboratory work, so you prepare a flagella stain of a culture while your friend prepares a wet mount. You cannot see any flagella on the cells, so claim the cells are not motile. However your friend claims to see motility in the wet mount. Could you have missed the flagella? Could your friend have been mistaken? Discuss the problems of the wet mount observations and flagella stain which can give conflicting observations.

References

Doetsch, R. N. 1981. *Determinative Methods of Light Microscopy.* In **Manual of Methods for General Bacteriology**. Gerhardt, P., editor. American Society for Microbiology. Washington, D.C. This reference covers some of the basic staining procedures used in this exercise, as well as other techniques applicable to the light microscope.

*The strain of *Klebsiella pneumoniae* recommended for use in this manual is ATCC strain #15574, which was isolated from plants and is unlikely to be pathogenic for humans. However, caution should still be observed.

Observation of Procaryotic Cells

Unknown code number _____

Fill in the chart on the observation of procaryotes

Characteristic	Proteus	Escherichia	Staphylo-coccus	Bacillus	Spirillum	Unknown
Cell morphology						
Cell arrangement						
Gram reaction						
Motility						
Distribution of flagella						

Tentative identification of unknown _____

Spore stain observations: Bacillus

Capsule stain observations: Klebsiella

(Continued on next page)

4

Observation of Eucaryotic Cells

Because of the generally larger size of eucaryotic cells, they are considerably easier to view microscopically than are procaryotic cells. In addition, internal structures can be recognized, and external structure is generally complex and distinctive. As a group, eucaryotes have less biochemical diversity than the procaryotes, but have considerably greater structural and morphological diversity. In this exercise, you will be observing some of the more common eucaryotic microbes in the groups of algae, protozoa and fungi. You are referred to your textbook for a detailed discussion of these three groups of eucaryotic microbes.

Algae and Protozoa Some of the most commonly encountered algae and protozoa are found in aquatic environments such as lakes and rivers. For this exercise, you are encouraged to obtain your own sample of lake or river water for examination. Since the density of cells in "open water" is fairly low, samples should be taken near rocks or other stationary objects to ensure sufficient cells for viewing. An interesting sample can be obtained by submerging a small jar right next to a rock or other fixed object which has a dense growth of algae (the wavy green material). Then, using your fingers or a stick, rub the deposit on the stationary object, allowing bits and pieces to flow into the jar. Such habitats are usually rich in algae, protozoa and bacteria. Since the species of organisms found in clean and polluted waters differ, it would be interesting for you to obtain samples from waters of various degrees of cleanliness. If there are no convenient rivers or lakes, look in drainage ditches, long-standing puddles and other sources of water.

Fungi The fungi you will be observing are selected from the types commonly found as contaminants from the air. Since, under rare circumstances, some of these fungi may cause disease, observe precautions when working with them. Healthy, young adults are very unlikely to be susceptible to any of these fungi, but once again, better safe than sorry. You might also wish to bring specimens of fungi from home (spoiled food from the back of a refrigerator for example). As you observe fungi, you will see that some of them grow in filaments, which are called **hyphae** (singular: hypha). These hyphae can be **septate**, where there are discernible cell walls between the individual cells, or **nonseptate**, where there are very few septa. The presence or absence of septa is one distinguishing feature used in the classification of the fungi. Masses of hyphae make up what is called a **mycelium** (plural: mycelia). Most fungi form spore-bearing structures. Notice the diversity of these spore-bearing structures and of the spores themselves. The spores observed are usually *asexual* spores, which serve as a major means of dissemination of the fungus. Refer to Figure 4-1 for the terms used to refer to some of these structures.

Materials

1. Water samples as described above obtained by the student or supplied by the instructor
2. Plate cultures of the filamentous fungi *Aspergillus carbonarius, Penicillium chrysogenum,* and *Rhizopus nigricans,* and the yeast *Saccharomyces cerevisiae*
3. Henrici slide demonstrations of the filamentous fungi listed in number 2. For a description of how these slides are prepared, see Appendix 2 on culture preparation.

Procedure

1. Prepare wet mounts of the water samples and look for the presence of algae and protozoa. You might also wish to try a hanging drop slide. To maintain the wet mounts for a long viewing period, seal the edges of the coverslip with vaseline, as directed in Student Supplement 1. (Confine the vaseline to the edges of the coverslip, so that it will not foul the objective lenses.)

    What do you see? Look for the algae based on their color and on the presence of internal structures. The most obvious internal structure will probably be the **chloroplast**, which contains the photosynthetic pigments. Many of the algal cells will be in long chains, but some will be single celled and possibly motile. Also look for **diatoms**, which are fragile-looking algae containing silica in their cell walls (see Figure 4-1). If your microscope is adjusted properly, you should be able to see the fine lines (striations) on the diatom cell walls. Now look for the protozoa. They should be very actively motile (and feeding!). Most of the protozoa you will see will probably be ciliates, characterized by the large number of cilia found around the periphery of the cells. The cells will lack the greenish coloration and the chloroplasts of the algae, but will usually have obvious food vacuoles and oral grooves. Watch the cilia around the oral groove as a protozoal cell feeds on bacteria. Amoebae might also be present, which move by sending out *pseudopodia* and feed by an engulfing process called *endocytosis*. If you have samples from both clean and polluted waters, compare the flora and fauna observed. Both of these sources might also contain metazoan organisms such as insect larvae, small crustaceans and molluscs. These organisms will be considerably larger than the microbial forms.

2. *Without opening the plates,* observe the colonial morphology of the cultures of filamentous fungi. Notice the coloration due to the spores. Turn the plates over to observe any water-soluble pigments exuded from the cells and to observe any disruption of the agar surface due to the growth of the fungi. Filamentous fungi such as these are usually called *molds.* The term "mold" has no taxonomic significance, but is used to describe any fuzzy-looking organism. You will become more familiar with molds

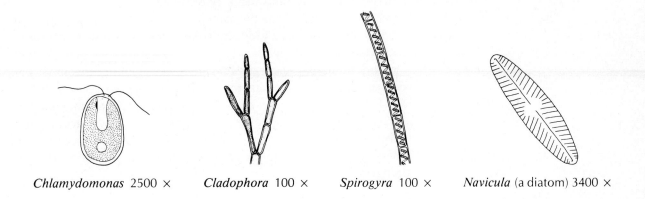

Chlamydomonas 2500 ×   *Cladophora* 100 ×   *Spirogyra* 100 ×   *Navicula* (a diatom) 3400 ×

Figure 4-1a   *Some algae commonly found in fresh water.*

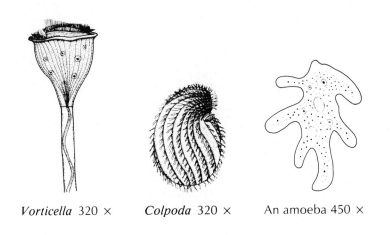

Vorticella 320 ×   *Colpoda* 320 ×   An amoeba 450 ×

Figure 4-1b   *Some protozoa commonly found in fresh water.*

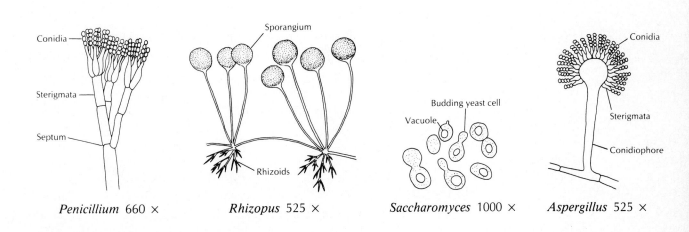

Conidia

Sterigmata

Septum

Penicillium 660 ×

Sporangium

Rhizoids

Rhizopus 525 ×

Budding yeast cell

Vacuole

Saccharomyces 1000 ×

Conidia

Sterigmata

Conidiophore

Aspergillus 525 ×

Figure 4-1c   *Common fungi.*

during the course as they will be frequently encountered as unwanted invaders on your agar plates.

The yeast culture, *Saccharomyces cerevisiae*, is not filamentous, does not form asexual spores, and has a colony morphology similar to that of bacteria. This organism is the same one (albeit a different strain) used in bread, beer and wine production, so it can be handled without fear of infection. Open the plate lid slightly. Can you smell an odor similar to that of home-baked bread? Prepare a wet mount of this fungus and compare what you see with the algae and protozoa seen earlier, and with the demonstration Henrici slides of the other filamentous fungi.

3. Observe the Henrici slides of the filamentous fungi. Pay special attention to the *substrate hyphae* on and embedded in the agar surface, the *aerial hyphae*, which are spore-bearing structures, and the spore structure and arrangement. Refer to Figure 4-1 for nomenclature.

Questions

1. What are the distinguishing characteristics of each of the three groups of eucaryotic microbes?

2. The habitat from which you obtained your algae and protozoa sample was probably very rich in microbial life. Referring, when necessary, to your textbook, who is eating whom? Keep this example of microbial interactions in mind as the course proceeds and you become more aware of the physiology of the microorganisms. Although you will mostly study isolated pure cultures of organisms, it is important to remember that, in their natural environment, microbes experience a variety of interactions with other life forms.

3. Define the following terms: hypha(e); mycelium(a); septum(a); conidiophore; conidium(a); sporangiophore; sporangium(a).

References

Raven, P. H. and R. F. Evert. 1981. **Biology of Plants**. 3rd edition. Worth Publishers. New York. An elementary text in botany which has good sections on both the algae and the fungi.

Vickerman, K. and F. E. G. Cox. 1967. **The Protozoa**. Houghton-Mifflin. Boston. Well-illustrated introduction to the protozoa.

Name _____

Date _____

Observation of Eucaryotic Cells

Observations of water samples

Observations of pure cultures of fungi

Culture	Colony morphology	Cell morphology (from Henrici slide observations or from a wet mount for *Saccharomyces*)
Aspergillus carbonarius		
Penicillium chrysogenum		

(Continued on next page)

(Continued)

Culture	Colony morphology	Cell morphology (from Henrici slide observations or from a wet mount for *Saccharomyces*)
Rhizopus nigricans		
Saccharomyces cerevisiae		

5

Growth

Growth can be defined as an increase in the amount of cell material. It is useful to distinguish between population growth (an increase in numbers or *biomass* of the population) and cellular growth (an increase in the size or biomass of an individual cell). Growth of microbes is usually defined in terms of population growth. The growth of a population of cells in a limited environment is often characterized by a growth curve with four phases: lag phase, log or exponential phase, stationary phase, and death phase. **Lag phase** begins when a population of nongrowing cells is supplied with a new source of nutrients. During this phase, the cells are metabolically active as they adjust to the new nutrient source, but cell division does not occur. The cells increase in size until they reach a critical size for division, at which point division begins, and the population moves into **exponential** phase. In this phase all cells are growing and dividing, and the progeny formed also can grow and divide. The cells are often also producing storage products from abundant nutrients in their environment. A great majority of microbes divide by a process called **binary fission**, in which each cell divides in two.* The increase in the population during this phase is generally thought of as two-fold exponential growth.

As the nutrient source is depleted or toxic waste products accumulate, the cells begin to alter their physiology: they might add on new layers of cell wall, slow down metabolism and division to match the lower nutrient availability of the environment, or begin to slowly degrade excess cellular materials and storage products. As the individual cells respond to the environmental change, the net effect on the population growth is the slowing of division, until no *net* division takes place. At this point, the population is in **stationary phase**. In this phase, if any cell growth occurs, it is matched by an equal amount of cell death. Depending on the type of organism, the stationary phase can be quite prolonged. Organisms which are normal inhabitants of rapidly changing environments such as soil often have evolved very efficient mechanisms for maintaining themselves during nutrient depletion. These mechanisms include, but are not limited to, the formation of resting structures, such as spores or cysts.

Eventually, the death in the population exceeds any cell division that might be occurring, and the population enters the **death phase**. The individual death of the cells could be due either to the starvation incurred as the cells run out of *endogenous reserves* and degrade some essential component, or to the toxicity of waste products. If **viable** cells in the death phase are supplied with new nutrients or are removed from the toxic environment, they may enter a new lag phase.

Measurement of cell growth

The measurement of cell growth can be accomplished by a number of methods. The advantages and disadvantages of each are discussed below.

Direct microscopic counts An increase in cell number in a population can easily be measured by counting the total number of cells (particles) in any given sample. This can be done microscopically by measuring the number of cells in a given volume of culture liquid. Special microscope slides are made for this purpose, such as the Petroff-Hausser counting chamber (for bacteria) or the hemocytometer (for blood cells and most eucaryotic cells). A major advantage of a particle count is the speed at which results are obtained. However, since it is often not possible to distinguish living from dead cells, the direct microscopic count method is not very useful for determining the number of viable cells in a culture.

Turbidity The cloudy appearance of a culture is called *turbidity*, and this property can be used to easily assess the amount of microbial growth. Turbidity can be measured by an instrument such as a colorimeter or spectrophotometer. These instruments contain a light source and a light detector (photocell) separated by the sample compartment. Turbid solutions such as cell cultures placed in the sample compartment interfere with light passage through the sample, so that less light hits the photocell than would if the cells were not there. Turbidimetric methods can be used as long as each individual cell can participate in light blocking; however, as soon as the mass of cells becomes so large that some cells effectively shield other cells from the light, the measurement is no longer accurate.

Turbidimetric measurements usually make use of a spectrophotometer which can be set up to use light at wavelengths of 600–700 nanometers (nm). Light in this range is used because it is rarely absorbed by cellular constituents, so any interruption of light passage is due to *reflection* not absorption. Using light at other wavelengths may introduce error, as both absorption and reflection are involved in light interruption.

To measure cell number using turbidimetry, you must experimentally derive a curve of turbidimetric measurement vs. cell number.

Viable counts A viable cell is defined as a cell which is able to divide and form a population. A viable count is usually done by diluting the culture, plating aliquots of the dilutions onto an appropriate culture medium, then incubating these plates under proper conditions so that colonies are formed. After incubation, the colonies are counted, and, from a knowledge of the dilution used, the original number of viable cells can be calculated (see Student Supplement 3). A major advantage of this method is that it is a direct measure of the viable cells in a population. It is also much more sensitive than the turbidimetric method. For accurate information, it is critical that each colony comes from only one cell, so chains and clumps of cells must be broken apart. However, since one is never sure that all such groups have been broken apart, the total number of viable cells is usually reported as **colony-forming units** rather than cell numbers. The major disadvantage of this technique is the time necessary for dilutions, platings, and incubations, as well as the time needed for media preparation.

Other methods of measurement Other measurement techniques are available when these commonly used techniques are inappropriate. For example, if one is interested in ascertaining the amount of some product produced by a culture so that the culture can be harvested when the product is at a high level, the product itself can be assayed by chemical assay methods. The amount of the product detected is an indirect measure of the growth of the culture and thus can be, with standardization, a measurement of growth. Still other methods, either direct or indirect, include measurement of substrate depletion or of the dry weight, nitrogen, or volume of the culture.

Approach of this exercise

For this exercise, you will be plotting cell growth measurements taken by turbidimetric techniques and viable plate counts obtained with a culture of procaryotic organisms.[†] The laboratory work of this exercise will include an introduction to pipetting and to viable cell counts, as well as to the mathematics of calculating generation times of microbes. You will not actually carry out the manipulations to generate a growth curve, but, as the course proceeds, you may be interested in performing a growth curve as a special project. Before beginning the laboratory exercise itself, it is advisable to have read over the explanation of dilution theory in Student Supplement 3 and to have worked the sample problems. You might also wish to practice pipetting a few times with a nonsterile pipette until you feel proficient. The major errors in this exercise are due to poor pipetting technique.

Period A

Materials

1. One to two ml of an *E. coli* culture of variable cell density, from 10^7 to 10^9 cells per ml[‡]
2. Two tubes containing 9.9 ml sterile water and four tubes containing 9.0 ml sterile water for dilutions
3. Fifteen sterile 1.0 ml pipettes
4. Ten plates of nutrient agar
5. Glass spreader and alcohol for sterilization
6. A sheet of linear and a sheet of 3-cycle semi-log graph paper

Procedure

1. Label two plates on the bottom for each of the following dilutions: 10^{-5}, 10^{-6}, 10^{-7}, 10^{-8}, and 10^{-9}.
2. Label the dilution blanks as follows: label the two tubes containing 9.9 ml sterile water as 10^{-2} and 10^{-4}; label the four tubes containing 9.0 ml sterile water as 10^{-5}, 10^{-6}, 10^{-7}, and 10^{-8}.
3. Carefully and aseptically remove exactly 0.1 ml of culture and pipette it into the tube marked 10^{-2}. Mix the tube completely, being careful not to spill any of the contents (if available, a vortex mixer is very useful for mixing dilution tubes). You have now prepared a 10^{-2} (1/100) di-

lution. Why is it 10^{-2}? 0.1 ml of undiluted culture was diluted into 9.9 ml of water, giving a total volume of 10.0 ml. $0.1/10.0 = 1/100 = 10^{-2}$. Do you follow this? If not, refer to the dilution problems in Student Supplement 3.

4. **Changing your pipette**, now take 0.1 ml from the 10^{-2} dilution and pipette it into the dilution tube marked 10^{-4}. Mix well. You have now prepared a *successive* 1/100 dilution, resulting in the total dilution of 10^{-4} (1/100 multiplied by $1/100 = 1/10,000 = 10^{-4}$).

5. **Change your pipette again**. Why keep changing pipettes? Because any fluid left in the pipette from the previous dilution will contain many more cells per ml than any successive dilution and, if used, will grossly confuse the final results by indicating a higher number of cells than were actually present in the original sample. (Refer to Student Supplement 3 again.) Now remove 1.0 ml from the 10^{-4} dilution, add it to the tube marked 10^{-5}, and mix well. What is your final dilution now?

6. **Changing your pipette between each tube**, continue your dilution series through 10^{-8}. If this is confusing, see Figure 5-1.

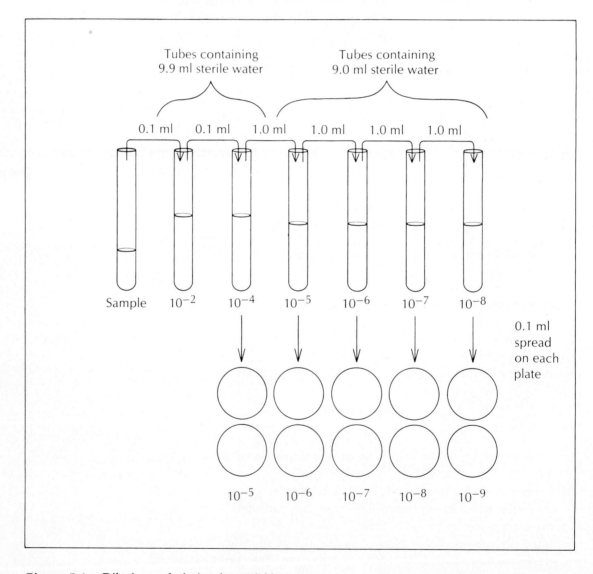

Figure 5-1 *Dilution and plating for a viable count.*

7. Spread a measured amount of a diluted culture on the appropriate agar plate. The spreading procedure is done in the following way. **Using a new pipette**, pipette 0.1 ml of the 10^{-8} dilution onto the agar of each of the plates marked 10^{-9}. Working quickly, dip the spreader into the alcohol, then touch it to the flame, allowing the alcohol to burn off. Do not keep the spreader in the flame but merely place it in the flame long enough to ignite the alcohol. Be careful that the burning alcohol does not drip onto your hands or papers. After the flame goes out, touch the spreader to a section of one of the plates that has no inoculum on it. Hear any sizzling? If so, your spreader was too hot. Proceed to spread the inoculum over the plate, turning the plate as needed to obtain complete coverage. Proceed to the second plate and spread that. Since these are *duplicate* plates of the same inoculum, you do not need to resterilize the spreader between plates. Why is this plate considered a 10^{-9} plate when you put inoculum from the 10^{-8} dilution tube into it? (Hint: final results will be given in colony forming units per ml, not per 0.1 ml.)

8. Continue to inoculate the other plates. To do this, use a new pipette for each dilution and resterilize the spreader for each set of duplicate plates. Remember that you will be pipetting 0.1 ml from each dilution onto *two* agar plates. The plates should be labelled as a ten-fold higher dilution than the dilution tube; for example, the 10^{-6} dilution should be plated on the agar plates marked 10^{-7}. Again, if you need help visualizing this, see Figure 5-1.**

9. Invert the plates and incubate them for 2 days at 37°.

10. In preparation for next period, graph the data given in Table 5-1. You should graph the number of cells vs. time on the 3-cycle semi-log graph paper, plotting time on the X-axis. On the linear paper, graph both the absorbance vs. time and the *logarithm* of the number of cells vs. time, plotting time on the X-axis again. On each graph, label the lag, log (exponential) and stationary phases. Also determine the rough slope of the line of the log of the number of cells vs. time plot. Refer to your text for a full explanation of the calculation of the generation time of bacterial growth before coming next period.

Safety Caution!

Table 5-1   Results of growth curve measurements of *Bacillus subtilis*

Time	Absorbance[1]	Cell Number[2]
9:30 a.m.	0.05	1.0×10^7
9:40	0.05	1.0×10^7
9:50	0.07	1.0×10^7
10:00	0.11	1.7×10^7
10:10	0.15	2.5×10^7
10:20	0.25	4.0×10^7
10:30	0.40	6.2×10^7
10:40	0.61	1.0×10^8
10:50	0.79	1.7×10^8
11:00	0.85	2.6×10^8
11:10	0.87	4.1×10^8
11:20	0.89	5.5×10^8
11:30	0.89	6.3×10^8

[1]Taken at 600 nm in a spectrophotometer.

[2]Determined by a viable plate count on nutrient agar; plates were incubated for 2 days at 37°.

Period B

Materials

1. Plates from the last period and, if possible, either a magnifying colony counter or a dissecting microscope
2. Graphs prepared last period

Procedure

1. Looking at your dilution plates prepared last period, choose the plates that have from 30–300 colonies on them. As this might take some practice in plate counting, you might need to choose all plates with what looks like a reasonable number of colonies to count. Using the colony counter or the dissecting microscope, count each colony. You will avoid counting a colony twice by marking off the colonies on the bottom of the plate as you count them. (This requires, of course, that the plate be upside down on the surface of the counter or microscope.) Be sure to count any small colonies. The advantage of performing these counts with magnification is the ability to see these small colonies, as well as the ability to detect "unusual" colony types, indicating different (contaminating?) colonies.

    Now look at the count of colony-forming units for the duplicate plates. Are the colony counts on the duplicate plates within 10% of each other?§ If so, your pipetting technique is good. Choose the results from the plates that really do have 30–300 colonies and determine the colony-forming units per ml of the original culture. How is this calculation made? See Student Supplement 3.

2. Observe your graphs. How do the arithmetic and log plots differ? What was your conclusion of the "real" lag phase on the linear paper: does it begin with the change in turbidity or the change in numbers? What was your result for the calculation of the generation time? The appropriate equation to use to calculate the growth rate constant, k, given in doublings per hour is:

$$K = \frac{\log_{10}N_t - \log_{10}N_o}{0.301\ t}$$

    N_o = Number of cells at initial time
    N_t = Number of cells at a later time, t
    t = Length of time from N_0 to N_t, expressed in hours.

Questions

1. What are the advantages and disadvantages of each growth measurement technique described above? Which would be most appropriate for a rough estimate of cell numbers of extremely virulent pathogens?
2. Why did you plate the dilutions *in duplicate*?

3. What is the generation time of the organism whose growth data are given in Table 5-1? Is this generation time fast or slow in comparison to other microorganisms? (To answer this, you will need to consult your textbook or discuss this with your instructor.)

4. With your knowledge of dilution theory, can you think of a way to perform turbidimetric measurements on very turbid suspensions of cells?

References

Cooper, T. 1977. *Spectrophotometry*. In **Tools of Biochemistry**. John Wiley and Sons. New York. A strong treatment of biochemical and physical basics of spectrophotometry and the spectrophotometer.

Koch, A. L. 1981. *Growth Measurement*. In **Manual of Methods for General Bacteriology**. Gerhardt, P., editor. American Society for Microbiology. Washington, D.C. Some elementary treatment of growth measurement techniques and a more advanced discussion of the statistics and calculations involved.

*This is not true of all cells, as some divide by budding or multiple fission. However these are exceptions.

†Actual growth curves are not generated because of the expense of supplies and the difficulty of performing so many dilutions with reasonable accuracy by a large group of students. Additionally, the availability of equipment might preclude the generation of these curves in some courses.

‡*E. coli* usually grows to a cell density of 10^9 overnight. To achieve a cell density of 10^7 or 10^8, dilute the overnight culture 1/100 or 1/10, respectively.

**If you work quickly enough, you can inoculate 4–8 of the plates without needing to stop to spread the inoculum. However, if you work too slowly, the inoculum may soak into the agar in the area of inoculation, making counting next period very difficult. If you do inoculate more than one dilution in a series, you need not change pipettes nor resterilize the spreader between dilutions *if you work from the most dilute (least concentrated) to least dilute (most concentrated) dilution*.

§To determine this, multiply one of the colony counts by 0.90 and 1.10 to get a span 10% above and below the original value. Compare the span of values to the second colony count. Does the second colony count differ by more than 10%?

Notes

Growth

Enumeration of bacteria

Dilution	Colony counts		Average
10^{-5}	_____	_____	_____
10^{-6}	_____	_____	_____
10^{-7}	_____	_____	_____
10^{-8}	_____	_____	_____
10^{-9}	_____	_____	_____

Number of colony-forming units/ml of original suspension: _____

Growth

Generation time in log phase of cells whose growth data is given in Table 5-1: _____

Notes

48

6

Environmental Parameters of Growth

The growth of an organism is affected by nutrient availability in its environment. However, presence of sufficient nutrients alone will not ensure growth since other parameters of the environment play important roles in determining the amount and rate of growth. In addition, other properties of an organism such as cultural morphology and pigment content can also be affected by the environment. Environmental factors which have significant effects on organisms include temperature, oxygen, pH, and osmotic pressure (water availability). Any of these parameters can deviate from the levels which would be optimal for a specific organism, resulting in a change in growth rate (even down to zero) and possibly causing a change in the morphology of the cells or of the colony.

In this exercise, the effect of each of these parameters will be studied. Although we will attempt to vary only one parameter at a time, you should be aware that changes in one parameter, especially extreme changes, can affect others. For example, as temperature or pH is varied, the solubility of oxygen also changes. In a natural environment, an organism must contend with all of the interactive changes that occur as changes in one variable start to affect the other variables.

Period A

The conditions in this exercise were chosen to represent ranges that might be reasonably expected in various habitats, and the organisms chosen are those which will be affected by changes in these conditions. However, for many of these organisms, their own natural environment is unlikely to show as much variation as they are subjected to here.

Materials

1. Two to three ml of one-day-old cultures of the following organisms:

   - *E. coli* — a normal inhabitant of the human intestines
   - *Pseudomonas fluorescens* — a normal inhabitant of soil and plant surfaces
   - *Bacillus subtilis* — a soil inhabitant
   - *Lactobacillus bulgaricus* — an organism commonly found on plant surfaces and in milk; often used in the production of yogurt; can also be found in the normal flora of some animals
   - *Streptomyces griseus* — a normal soil inhabitant, produces geosmin (the earthy smell of soil) and some antibiotics
   - *Clostridium butyricum* — a common soil anaerobe
   - *Leuconostoc mesenteroides* — an organism also found on plant surfaces; involved with the production of sauerkraut
   - *Saccharomyces cerevisiae* — a yeast (eucaryote) found on plant surfaces and used to produce bread, beer, wine and other fermented foods

   Note that the organisms supplied are generally normal inhabitants of soil and plant environments. We have purposely omitted the use of human pathogens in this exercise, as we have throughout the manual. However, it would be useful for you to think of how you would expect human pathogens to react to these environmental changes.

2. For the oxygen experiment, five tubes of thioglycollate agar, melted and in a 45° waterbath

3. For the temperature experiment, four plates of nutrient agar (NA)

4. For the experiment on osmotic pressure, one plate each of NA, NA + 5% sucrose, NA + 10% sucrose, NA + 20% sucrose

5. For the pH experiment, one plate each of yeast extract-glucose (YEG) agar adjusted to pH values of 4, 5, 6, 7, and 9

Procedure

1. Obtain the five tubes of thioglycollate agar from the 45° waterbath. You should probably bring them to your desk in a beaker (can, glass) of 45° water either from the water bath or from a tap (hot tap water is usually at least 45°; the temperature should be adjusted to 45° using cold water). Prepare a second beaker of cold water to receive the tubes after inoculation. Label a tube for each of the following cultures: *E. coli, P. fluorescens, L. bulgaricus, C. butyricum,* and uninoculated control. Working rapidly (the tubes will solidify when they cool!), inoculate the appropriately labelled tube with a loopful of culture. Be sure to insert the loop to the bottom of the tube and, after inoculation, to roll the tube in your hands to distribute the inoculum in the medium. After each inoculation, rapidly cool the inoculated tube in the beaker of cool water. (Why cool rapidly? Some of these organisms may be injured if left for long at the high temperature at which the agar is kept.) Incubate the tubes at 30° for 2 days.

2. On the bottom of each of the four NA plates to be used for the temperature experiment, draw lines to divide each plate into three pie-

shaped sections. Label one section on each plate for *E. coli*, one for *P. fluorescens*, and one for *B. subtilis*. Using the loop, spot inoculate each sector with the appropriate culture. Allow the loopful of material to soak into the agar before inverting the plate for incubation. Incubate one plate at each of the following temperatures: 10°, 24°, 37°, and 45°. Make observations after two days of incubation.

3. Divide each plate of YEG agar into five sections and label each section for one of the following cultures: *E. coli, P. fluorescens, L. bulgaricus, S. griseus,* and *S. cerevisiae.* Using the loop, inoculate each culture in the appropriate section of each plate. The inoculation can be a spot or a line, but be careful that the inocula do not run together. After the inocula have soaked into the agar, invert the plates and incubate them at 30° for two days.

4. In a similar manner, divide and label the NA + sucrose plates for the following *three* cultures: *P. fluorescens, L. mesenteroides,* and *S. cerevisiae.* Inoculate the cultures as in #3 above. Incubate the plates at 30° for two days.

Period B

Materials

Plates and tubes inoculated last period

Procedure

1. Observe the thioglycollate tubes for amount and position of growth. Use the uninoculated control to determine if growth occurred in the tube. Record your results on Report Sheet 6. Compare the results with those in Figure 6-1 and attempt to classify the organisms you used.
2. Rate the amount of growth on the plates incubated at different temperatures. To do this, make a simple qualitative estimate of amount of growth, such as − (no growth), 1+ (small amount of growth), 2+, 3+, and 4+ (increasing amounts of growth). Record the results.
3. Observe the plates from the pH experiment for growth (presence and amount). Also note any differences in colony morphology or color.
4. Observe the plates from the sucrose experiment for both amount and type of growth. Be sure to note any morphological differences in the colonies as the percentage of sucrose increases. Record the results.

Questions

1. What could account for the greater amount of growth at the top of the thioglycollate tube inoculated with a facultative anaerobe? (You may wish to read about respiration and fermentation in your textbook.)

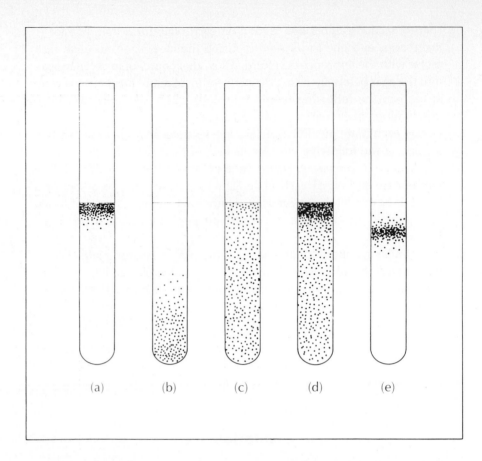

Figure 6-1 *Oxygen requirements as shown by growth responses of bacteria cultured in thioglycollate agar tubes: (a) aerobe—growth only at top of tube; (b) obligate anaerobe—growth only at bottom of tube; (c) aerotolerant anaerobe—even growth throughout; (d) facultative anaerobe—best growth at top of tube, but growth throughout; (e) microaerophile—growth near, but not at, surface and no growth below.*

2. What prevents a strict anaerobe from growing at the top of the thioglycollate tube?

3. What component(s) of the thioglycollate medium help keep oxygen levels low in the bottom of the tube? What component(s) of the medium indicate the presence of oxygen to you? (To answer these questions, you will need to refer to Student Supplement 2, which has a list of ingredients of each medium and a discussion of the purpose of important components.)

4. If there is no growth of one of the organisms at 45° or 10° after two days, can you conclusively say that the organism is unable to grow at this temperature? If not, why not?

5. After complete testing, you have decided that Organism A cannot grow at either 55° or 4°. Give a simple biochemical or physiological explanation of inability to grow at each of these temperatures. How could you test to determine whether the cells had been killed at the specified temperature or had just not been able to grow?

6. Look at the plates of YEG that are at different pH values. Is there any indication that pH might *not* be the sole factor determining the growth of an organism at a specific pH? (Look at the medium itself. Any change of color? Precipitate?)

7. What might explain the inability of an organism to grow on the nutrient

agar plate that had *no* sucrose added to it? (To answer this, it might be useful to look at the growth characteristics of that organism in the thioglycollate tube. What sort of organism is this and how does it obtain its energy for growth? Think about fermentable and nonfermentable energy sources.)

8. What would you expect the growth results to be in the experiment on osmotic pressure if you substituted glucose for the sucrose in the agar plates? Consider both presence and type of colonies.

References

Costilow, R. N. 1981. *Biophysical Factors in Growth.* In **Manual of Methods for General Bacteriology**. Gerhardt, P., editor. American Society for Microbiology. Washington, D.C. A useful reference on the physical factors affecting growth.

Notes

54

Control of Microbial Growth

7

In the previous exercise, you saw how environmental parameters can affect microbial growth. Manipulation of these parameters can be used to control microbial growth. In addition, specific control measures can be used to kill or to inhibit the growth of microorganisms. A procedure which leads to the death of cells is broadly termed **cidal**, whereas, a procedure which only inhibits growth is termed **static**. If the organism being killed is a bacterium, the term used is **bactericidal**; killing of a fungus would be referred to as **fungicidal**, and so forth. Cidal measures are used when it is important that living microorganisms be completely eliminated, for example, when canning food, preparing media for laboratory use, or wiping a laboratory bench with disinfectant. Static measures are used when the organisms need not be killed, but when their growth must be slowed or stopped. An example of static control is the use of refrigeration to slow microbial growth in foods, thus increasing their storage life.

Many chemicals are used to control microbial growth. The modes of action of growth-controlling chemicals vary, but the most common actions are disruption of cell membranes or interference with nucleic acid or protein synthesis. Some of the most commonly used chemicals are classified as antiseptics, disinfectants, or antibiotics. **Antiseptics** are chemicals which kill or inhibit the growth of microorganisms and are safe to use on animal tissue. Thus, antiseptics may be used to treat wounds or as mouth washes. **Disinfectants** are chemicals able to kill microorganisms which are generally not safe to use on animal tissue. Disinfectants find wide use in the home, laboratory, hospital, and food industries to disinfect surfaces.* **Antibiotics** are chemicals produced by one microorganism which kill or inhibit the growth of another. Antibiotics are used both internally and externally in the treatment of disease. Antibiotics will not be studied here but will be studied in Exercises 17 and 21.

In this exercise, you will study the effects of various antiseptics and disinfectants on cell viability and growth. Keep in mind that chemical treatment is only one of many methods available to control microbial growth. Other methods include cidal control measures such as sterilization by heat (e.g. autoclaving) and by use of ultraviolet radiation, and static control measures such as freezing and drying. The method of choice depends on the material to be treated, the requirement for cell death vs. growth inhibition, the organism to be controlled, and characteristics of the method such as ease of use, danger to the user, and cost.

Period A

In this period, you will testing the effectiveness of a variety of antiseptics and disinfectants on bacterial growth. If you wish to contribute your own commercial product for testing, be sure to bring it in for the instructor to dilute before class. Since it is impossible for every student to test every chemical, you should plan to discuss your results with other students in Period B.

Materials

1. One to two ml of overnight cultures of *E. coli* and *S. epidermidis*
2. A dilution series of various antiseptics and disinfectants. A recommended dilution series is: undiluted; 1/5; 1/10; 1/20; 1/40; and 1/80. Antiseptics and disinfectants used should include the bench disinfectant used in your laboratory, commercial disinfectants, commercial antiseptics including mouthwashes.
3. Six small sterile filter paper disks to soak with antiseptic or disinfectant
4. Two plates of nutrient agar (NA)
5. One pair of forceps and alcohol to flame the forceps
6. Two tubes of melted top agar cooled to 45–50° in a waterbath
7. Two sterile 1 ml pipettes

Procedure

1. Label one NA plate for *E. coli* and the other for *S. epidermidis*.
2. Remove one tube of melted and cooled top agar from the waterbath. Wipe off the outside of the tube with a paper towel (this removes the contaminated water which might otherwise drip into the plate when you pour the agar). Aseptically add 0.1 ml of *E. coli* to the tube, mix gently but thoroughly, then pour the top agar into the nutrient agar plate. Quickly tilt the plate to spread the agar overlay completely over the agar surface. Set the plate aside to solidify for 5–10 minutes before starting step 3.

   Repeat this procedure with the culture of *S. epidermidis* and the other plate. You have now prepared plates in which the bacteria will grow in the thin overlay giving a turbid appearance to the surface of the plate. Growth inhibition will be visible as a clear area in this "lawn."
3. Divide each plate into sectors by labelling the bottom of each plate to accommodate the various dilutions of your assigned antiseptic or disinfectant. Be sure to record which chemical you are using.
4. Flame-sterilize the forceps by dipping it into the alcohol, then touching it to the flame. Allow the alcohol to burn off. It is *not* necessary to continue holding the forceps in the flame; you are not sterilizing the forceps in the same way you do a loop. Be careful to hold the forceps in such a way that the burning alcohol does not drip down the handle to your hand, nor drop onto your notebook. Done carefully, flame sterilization of forceps and other equipment will not result in fires or burns.

Safety Caution!

Now pick up a sterile filter paper disk with the forceps. Proceed from the least concentrated to the most concentrated solution. Saturate the disk with the liquid of your antiseptic or disinfectant. Allow excess liquid to drain off by holding the disk against the side of the tube or bottle, then place the disk in the appropriate area of the *E. coli* plate. Prepare disks for the other dilutions, *working from the most dilute to the least dilute concentration*. Resterilize the forceps and repeat this procedure to apply disks to the *S. epidermidis* plate. Invert and incubate the plates at 37° for 2–5 days. (The disks will not fall into the lid of the plate, as hydrostatic forces, water's attraction for itself, are very strong!)

Period B

Materials

1. All the plates prepared in period B
2. A millimeter ruler

Procedure

Observe the plates containing disks of your chemical. Measure the **zone of inhibition** (clear area) around each disk, using a ruler placed on the bottom of the plate. How effective is your chemical in controlling growth? What is the highest dilution that is still effective? Observe results of other students. If you see any changes in the agar (such as color change or precipitation) around the disk of chemical, could the effect on the growth of the organism be due to a change in the nutrient availability rather than the effect of the chemical on the cell itself?

Questions

1. Why is it important that the agar overlay be at 45–50° before adding the bacterial cells?
2. How can you determine if the organisms in the zone of inhibition have been killed or just inhibited? (This brings up the question of survival time of organisms that are not growing: Many organisms can survive for a considerable period of time in the presence of growth-inhibiting chemicals. The cells will slowly metabolize either endogenous or exogenous sources of energy until the inhibiting chemical is removed. However, some organisms are not as well adapted to this long-term survival. To answer the question posed, assume these cells are capable of long-term survival.)
3. Is the zone of inhibition an absolute measure of the effectiveness of the compound in controlling microbial growth? To answer this, consider the effect of the rate of diffusion of the compound through the medium, the

length of time of exposure to the chemical, how the chemical might be used in actual practice, and the physiological state of the organisms tested (i.e. are organisms more resistant to the effect of chemical inhibitors when they are in log phase? stationary phase?).

4. Think of situations in everyday life in which various measures are used to control microbial growth. Consider the areas of health, food, sanitation, and material preservation (e.g. telephone poles).

References

Barkley, W. E. 1981. *Containment and Disinfection.* In **Manual of Methods for General Bacteriology**. Gerhardt, P., editor. American Society for Microbiology. Washington, D.C. This article has a small section on disinfectants which raises pertinent questions on their proper use.

*The term disinfectant is sometimes used when describing chemicals such as iodine solutions (common trademark: Betadine) used on the skin in preparation for surgery. However, most disinfectants are not used on animal tissue.

Control of Microbial Growth

Effectiveness of disinfectants and antiseptics

Disinfectant or antiseptic used: _____

Dilution	Zone of inhibition in mm	
	Escherichia coli	*Staphylococcus epidermidis*
_____	_____	_____
_____	_____	_____
_____	_____	_____
_____	_____	_____
_____	_____	_____

After observing other students' data, which disinfectant and which antiseptic are most effective against

E. coli _____

S. epidermidis _____

Notes

8

Biosynthesis and Nutrition: Catabolism

In order to grow and divide, the cell must take in nutrients from its environment, process these nutrients to release energy and form building blocks, polymerize these building blocks into macromolecules, and excrete waste products. All of these activities together are called **metabolism**. It is often useful to divide metabolism into two parts: degradative processes, collectively termed **catabolism**, and biosynthetic processes, collectively termed **anabolism**. We have used this division of metabolism to devise laboratory exercises of convenient length. However, you are reminded that this division often obscures the very real and necessary interrelationship between the two groups of processes. We will attempt to bridge this gap with pertinent questions at the end of the exercises.

Catabolic processes are involved in the provision of energy for cellular activities such as biosynthesis, transport, and motility. For *heterotrophic organisms* (organotrophs), catabolic processes involve the degradation of organic materials to yield energy and waste materials. In many cases, catabolism also results in the formation of small organic molecules that serve as carbon skeletons for biosynthesis. The ability of a particular organism to use a specific compound or form specific waste products is determined by the genetic information the organism carries. Therefore, the determination of the activity of an organism on any compound can be used in taxonomic identification. (After all, that is the essential basis of taxonomy: genetic similarity.) In this exercise, you will observe some of the differences between heterotrophic bacteria in terms of what compounds they can use for energy production and what metabolic waste products they make. For a more complete discussion of the biochemistry involved, refer to your textbook.

Period A

In this period, you will be inoculating various media with different organisms. Read over the constituents of the media and their purposes in Student Supplement 2. Before next period, be sure to read the pertinent material in Student Supplement 1 describing the chemical tests you will be performing to detect endproducts.

Materials

1. Two to three ml of overnight cultures of *E. coli, Klebsiella pneumoniae, S. epidermidis,* and *M. luteus*
2. Five tubes of tryptone broth
3. Five tubes of methyl red-Voges-Proskauer (MRVP) broth
4. Five glucose Durham tubes, five sucrose Durham tubes, and five lactose Durham tubes, all containing bromcresol purple as a pH indicator
5. Three plates of mineral salts medium (MS) each containing 0.05% yeast extract and 0.5% of one of the following sugars: glucose, sucrose or lactose

Procedure

1. Label one tube of each medium for each culture (the fifth tube of each is to be used as an uninoculated control). Divide each plate into sectors and label one sector on each plate for each organism.
2. Using a loop, inoculate one of each type of tubed medium with each culture. You should end up with 5 different tubes inoculated with each culture. If your aseptic technique is good, there is no need to resterilize the loop after inoculating one tube and before recharging your loop with culture for the next inoculation. (An alternative to this is to use a pipette for inoculating the tubes.) Gently roll each tube to distribute the inoculum in the tube. This is especially important for the Durham tubes in order to get some of the inoculum into the small inverted tubes. The purpose of the small inverted tube is to trap gas. Be careful not to roll the tube so hard as to get air into this small tube.
   Incubate the tubes at 37° for 2 days.
3. Spot inoculate each of the MS + YE + sugar plates with each of the cultures (resulting in each of the four cultures on each plate). Allow the inoculum to dry, then invert and incubate the plates at 37° for 2 days.

In this period, you will be making observations of some of the inoculated media from last period. You will then save the media to observe again in another 2 to 5 days, since some reactions may not be complete.

Materials

1. Tubes and plates inoculated in Period A
2. Demonstration tubes of glucose agar deeps, lactose agar deeps, and sucrose agar deeps inoculated with each of the four cultures used in this exercise. Before inoculation, these tubes were heated to melt the agar and cooled to 45°. Inoculation was performed just as the thioglycollate tubes were inoculated in Exercise 6.

Procedure

1. Observe the Durham tubes for acid and gas. Acid is indicated by the change in color of the bromcresol purple dye from purple to yellow. This dye is a pH indicator, as are some other dyes you will see in later laboratory exercises.* If the dye is yellow, the organism has *fermented* the sugar to acidic endproducts. Examples of pathways by which an organism might do this can be found in your textbook. Depending on the pathway used, gases such as hydrogen and carbon dioxide might be endproducts. Gas is indicated by the presence of a bubble (the absence of liquid) in the Durham tube. If you are unsure about the presence of gas, use the control tube for comparison. Record your results on Report Sheet 8.
   Most organisms that can ferment sucrose or lactose can also ferment glucose. This is because in order to metabolize efficiently, an organism will generally funnel a complex compound into a pathway for degradation of a simpler compound. This is what happens in the cases of sucrose and lactose. Sucrose is a *disaccharide* made up of the monosaccharides glucose and fructose. Organisms that ferment sucrose have an enzyme (sucrase) that cleaves the bond between the glucose and fructose moieties. The glucose is then fermented by the same pathway the organism would use for exogenously supplied glucose. Fructose is an isomer of glucose and can be easily changed by the cell into glucose. Similarly, lactose is a disaccharide composed of glucose and galactose. The cell can produce an enzyme (lactase, more commonly called β-galactosidase) that cleaves the β-galactoside linkage between glucose and galactose. Glucose is then funneled into the normal glucose pathway, and galactose is chemically modified to feed into similar pathways.
2. Observe the demonstrations of agar deeps. What would be the advantages and disadvantages of using agar deeps instead of Durham tubes?
3. Observe the MS plates for growth. Record your results. What is the purpose of using these plates? Would the Durham tubes give you all

the information you might need about an organism's utilization of a particular carbon and energy source?

4. Reincubate all tubes and plates for another 2 to 5 days.

Period C

Materials

1. Tubes and plates from Period A and B
2. Dropper bottles of reagents for the methyl red test, the Voges-Proskauer test for acetoin, and the indole test. The formulas and reactions for these reagents are given in Student Supplements 1 and 2.
3. Five empty tubes

Procedure

1. Observe the Durham tubes for acid and gas. Be sure to note any changes from the 2-day observations. Record the results. In some cases, tubes that were basic (purple) after two days incubation may now be acidic (yellow). Explain. In other cases, the tubes might have been acidic after two days, but basic after 4 to 7 days. The cause of this may be less obvious to you, but there are several possibilities, each of which requires that the sugar substrate be depleted. After the sugar is gone, the cells must use an alternative substrate for their energy source. One possibility is the utilization of the acidic endproducts produced earlier. Although these are endproducts of *fermentation*, it is possible for some organisms to use these as substrates for *respiration*. Respiration involves the use of oxygen as a terminal electron acceptor, and the endproduct of aerobic oxidation is CO_2, which is less acidic than the fermentation endproducts. Why would the system turn from anaerobic to aerobic? In a culture containing a facultative organism, anaerobic conditions are brought about by the organism itself as a result of the consumption of oxygen during the oxidation of the easily degraded sugar substrate. Once the sugar substrate is depleted, the oxygen consumption rate becomes reduced and the culture becomes reaerated via diffusion of oxygen from the atmosphere. This reaeration allows the organism to respire, and the color change from yellow back to purple (as the organism uses up acidic endproducts of fermentation) occurs near the surface of the broth first. Other possible substrates are amino acids supplied by the tryptone or yeast extract. In this case, the oxidation of amino acids leads to the production of ammonia, which is an alkaline product. This process is often only possible if the organism has an adequate oxygen supply for respiration, so the resulting pH change appears similar to that seen when the acidic endproducts are catabolized (at the surface of the broth first). Notice, however, that the acidic endproducts in this case are still present, albeit in a neutralized chemical form.

66

2. Observe the plates containing MS + YE + sugar for growth. Since the medium contains too little yeast extract to support extensive growth of the inoculated bacteria, any good growth must be due to the utilization of the sugar as an energy source. Compare the results with those obtained for each organism in the Durham tubes. Can you see any purpose in testing the ability to use a substrate with both methods? (Put another way, does the absence of acid and gas in the Durham tube indicate *inability* to use the supplied sugar? Conversely, is growth in the Durham tube without the presence of acid and gas sufficient proof of the *ability* of an organism to use the sugar supplied?)

3. Test each tryptone broth tube for the presence of indole by adding several drops of Kovac's reagent to the broth. Do not mix. Any indole will react with the reagent to form a red ring at the surface of the broth. Indole is produced as a waste product in the cleavage of tryptophan by the enzyme tryptophanase, yielding indole and pyruvate. The pyruvate enters the central metabolic pathways of the cell and is used for energy and biosynthesis. The ability of an organism to produce the tryptophanase enzyme is, of course, genetically determined, so the production of indole under the proper conditions can be used to differentiate bacteria taxonomically.

4. Label each of the empty tubes to correspond to the cultures used in this exercise. One tube will be labelled control. Split the cultures grown in MRVP broth into equal amounts between the original culture tube and the new empty tube. This can be accomplished by pouring the culture or by pipetting. (If a pathogen were involved, pipetting would be the *only* way! Why?) Use one tube of each culture for the methyl red test and the other for the Voges-Proskauer test.

    The methyl red test is performed by adding a few drops of methyl red to the broth. If the broth turns red, the pH is below 4.6. This is considered a *positive* methyl red test. If the broth stays its original color, the pH is above 4.6 and the test is considered *negative*. The final pH of the broth is indicative of the pathway an organism has used in the fermentation of the glucose in the medium. Some organisms use a pathway called the mixed-acid pathway, where the endproducts are all acidic. These organisms are typically methyl-red positive. Other organisms might use pathways that produce neutral endproducts as well as, or in place of, the acidic products. These organisms are generally methyl-red negative. Two neutral endproducts produced by a wide variety of organisms are acetylmethylcarbinol (acetoin) and butanediol. These endproducts can be detected by the Voges-Proskauer test.

    The Voges-Proskauer test is performed by adding one drop of α-naphthol to the broth, mixing, and then adding 10 drops of 40% KOH. Since aeration increases color development, shake the tube to aerate. The development of a pink color within 15 to 20 minutes is an indication of a positive test, i.e. presence of either acetoin or butanediol. The production of acetoin is very important in butter production, as this compound gives butter some of its characteristic flavor.

Questions

1. Devise a possible plate method to test for *both* utilization and fermentation of a sugar.

   When testing sugars, it is often desirable to get a small amount of growth even when the principle sugar is not utilized just to prove that the inoculum was successfully transferred and that the cells were viable. This prevents false negative results. Does your medium allow for this?

2. Which organism would you expect to be more acid-tolerant and why: *E. coli* or *K. pneumoniae*?

3. Given that carbon dioxide gas is very soluble in alkaline solutions and hydrogen gas is not, can you think of any way to determine the composition of gases in the Durham tube? Could use of your test suggest another possible advantage of using Durham tubes over agar deeps?

4. If color change in the Durham tubes due to deamination of amino acids is a potential problem, could the medium be modified so that it would be more like the medium containing mineral salts and low yeast extract? (Hint: Be aware that the Durham tube fermentation test is used diagnostically for a very wide range of bacteria. You might want to return to this question after the section on growth factors in the next exercise.)

References

Lehninger, A. L. 1980. **Biochemistry**. 3rd edition. Worth Publishers, Inc. New York. A good biochemistry text outlining the metabolic pathways used by various organisms.

Smibert, R. M. and **N. R. Krieg**. 1981. *General Characterization*. In **Manual of Methods for General Bacteriology**. Gerhardt, P., editor. American Society for Microbiology. Washington, D.C. Detailed explanation of a number of different laboratory tests, although rarely discusses the biochemical bases of the reactions.

*However, not all dyes change color with pH.

Biosynthesis and Nutrition: Catabolism

Fill in the spaces in the following chart with a + or −

Test	Escherichia coli 2 day	Escherichia coli 5 day	Klebsiella pneumoniae 2 day	Klebsiella pneumoniae 5 day	Staphylococcus epidermidis 2 day	Staphylococcus epidermidis 5 day	Micrococcus luteus 2 day	Micrococcus luteus 5 day
Glucose fermentation: acid gas								
Glucose utilization								
Lactose fermentation: acid gas								
Lactose utilization								
Sucrose fermentation: acid gas								
Sucrose utilization								
Tryptophanase (indole test)	✕		✕		✕		✕	
Methyl red test	✕		✕		✕		✕	
Voges-Proskauer test	✕		✕		✕		✕	

Which of the above four organisms are likely to be aerobes?

Which of the above four organisms are likely to be facultative anaerobes?

(Continued on next page)

Which of the above four organisms use a neutral-endproducts pathway in the fermentation of glucose?

Which of the above four organisms might be able to use tryptophan as the sole source of carbon and energy for growth?

Biosynthesis and Nutrition: Anabolism

In the last exercise, the use of different substrates as energy sources for heterotrophic organisms was studied. For most of these organisms, the organic energy source also provides carbon skeletons for biosynthesis of monomers (e.g. amino acids, nucleic acids) and polymers (e.g. proteins, nucleic acids, peptidoglycan). To make these monomers and polymers, the organism must make a wide variety of enzymes, most of which require cofactors and/or coenzymes. Cofactors are usually cations such as potassium, magnesium, or calcium, whereas coenzymes are usually low-molecular-weight organic molecules, such as vitamins. Although cofactors must be obtained by the organism from the environment, coenzymes are often synthesized by the organism as part of its biosynthetic processes. However, some organisms lack the genetic information to make some of their required coenzymes or even to make some of the monomers required for polymer synthesis. These organisms must then have these small organic molecules supplied to them in their environment. Low-molecular-weight organic molecules required by a cell *in small amounts* are called **growth factors**. The inorganic compounds such as the cations mentioned above are required nutrients but are not considered growth factors.

In addition to the need for specific inorganic and organic compounds for enzyme activity, the organism also requires a supply of elements such as sulfur, nitrogen, and iron as integral parts of cellular constituents. For example, sulfur and nitrogen are both elemental constituents of amino acids, and iron is an integral part of the cytochromes used by respiring cells. Organisms must not only have a supply of these elements, but the elements must be in the correct chemical form for utilization. Nitrogen (as N_2) makes up about 80% of the atmosphere of earth. Most organisms cannot obtain their nitrogen in this chemical form, however, so need it supplied as the ammonium (NH_4^+) or nitrate (NO_3^-) ion.

To study the requirement of different nutrients and growth factors, you will be inoculating different cultures onto plates containing different agar media. For many of the elements required in small amounts, it is very difficult to prepare culture media lacking the element, because of chemical contaminants in the other ingredients added to the medium. For this reason, only those elements required in large amounts will be studied.

Period A

In this period, you will be inoculating a variety of media containing different growth factors. To minimize carry-over of nutrients from one medium to another, you should inoculate the plates *in the order specified*.

Materials

1. One to two ml of cultures of wild-type *E. coli*, an *E. coli* strain auxotrophic* for an amino acid†, and *Lactobacillus bulgaricus*
2. One plate of each of the following media: minimal medium number 1 (MM1), MM2, MM3, MM4, and all purpose Tween (APT) agar. The ingredients for these media are given in Student Supplement 2.

Procedure

1. Divide the plate into three sectors by marking with a wax pencil and label each plate for the inoculation of all three cultures.
2. Inoculate each culture onto the agar in the appropriate sector of each plate using your loop. The inoculation can either be a spot or a line, but be sure to recharge the loop with inoculum between each plate. Inoculate the plates in the order: MM1, MM2, MM3, MM4, APT agar. Allow the liquid to dry thoroughly before inverting the plates, to prevent the inocula from running together.
3. Incubate the plates at 30° for 2–5 days.
4. Before next period, look at the formulations for each of these media. Try to predict the pattern of growth you should expect for each organism. Look up each organism in your textbook or in Bergey's Manual (reference given below) to determine its growth requirements.

Period B

Materials

Plates inoculated during the last period.

Procedure

Observe each plate for growth for each organism. Record your observations on Report Sheet 9. Rate the amount of growth of the organism on each plate by comparing it to the best growth you see for that organism on any plate.

Questions

1. Looking first at wild-type *E. coli*, which plates could it grow on? Which could it *not* grow on? What is missing from the plates it could *not* grow on? (Help? Look in your textbook in the section on nutritional requirements for growth. Pay special attention to which elements are needed in fairly large quantities, and why.)
2. Next, look at the mutant *E. coli* which is an *auxotroph*. What medium(a) will *not* support its growth? Why not?
3. Define and give an example of a growth factor.
4. Now, look at the *Lactobacillus* culture. Could it grow on as wide an array of media as *E. coli*? Why not? To answer this, remember that the ability of an organism to use certain substrates or its requirement for certain growth factors is determined by the genetic information it carries. If an organism lacks the enzyme to degrade lactose, it will not be able to use lactose for carbon or energy.
5. Look up the definition of the word "fastidious" in your textbook. Would you expect to find a fastidious organism in an environment poor in organic materials?
6. What would be the effect on the growth of wild-type *E. coli* of omitting the glucose from MM1? (Recall that *E. coli* is a heterotroph.) Could anything else be substituted for the glucose in the medium (go back to the last exercise to answer this)?
7. What is the missing component in MM2? What is it used for? Is there any substitute for this component?
8. Now return to the previous exercise to answer the question about the ingredients for the medium in the sugar Durham tubes.

References

Buchanan, R. E. and N. E. Gibbons, editors. 1974. **Bergey's Manual of Determinative Bacteriology**. 8th edition. Williams and Wilkins. Baltimore. Good source of characteristics of specific bacteria.

Guirard, B. M. and E. E. Snell. 1981. *Biochemical Factors in Growth*. In **Manual of Methods for General Bacteriology**. Gerhardt, P., editor. American Society for Microbiology. Washington, D.C. A reasonably complete overview of cellular nutrient requirements.

*An auxotrophic mutant is one which requires a growth factor which the wild-type does not require. A wild-type is a strain of organism that is normally isolated directly from the wild habitat, which in the case of *E. coli* is the animal intestinal tract.

†Any amino acid auxotroph of *E. coli* can be used for this exercise. The required amino acid should be added to the media as indicated in the media formulations given in Student Supplement 2.

Notes

74

Biosynthesis and Nutrition: Anabolism

Fill in the following chart indicating growth from − for no growth to + for best growth.

Medium	Escherichia coli wild type	Escherichia coli auxotroph	Lactobacillus bulgaricus
All purpose Tween agar			
Minimal medium 1			
Minimal medium 2			
Minimal medium 3			
Minimal medium 4			

If there is no growth on any of the above media, explain.

Notes

10

Culture Media

In the last few exercises you have used a wide variety of media to test for growth requirements. You have been instructed to study the medium constituents for information about the nutritional versatility of organisms. It is now appropriate to discuss the *classes* of culture media that are used in microbiology. Culture media can be classified chemically, nutritionally, or functionally. *Chemically,* a medium can be either **defined** or **undefined (complex)**. In a defined medium, the user knows all the components used to make up the medium. For example, a medium containing 5 grams of glucose, 1 gram of $(NH_4)_2SO_4$, and 2 grams of KH_2PO_4 is a defined medium, since all the constituents are known. Keep in mind that this defined medium has a small amount of chemical contaminants, like iron or chloride, from the chemicals and water used. The contaminants introduced from the chemicals themselves are usually identified on the label by the manufacturer, but the contaminants from the water are not known, unless an analysis of the water is performed. However, since distilled water is usually used for media preparation, the level of chemical contaminants from the water is relatively low. What if an amino acid were added to the medium listed above? Is the medium still a defined one? Yes. Addition of any *known* component does not change the status of the medium. However, if a complex component like beef extract, yeast extract, peptone, or other plant or animal extracts were added, the medium would be called *undefined* or *complex*. The reason for this is that the chemical make up of these extracts is not completely known, so that unknown chemicals are introduced into the medium. In many cases, such as when cells are being grown for inoculation into other media, the presence of unknown compounds is of little consequence. However, when specific biochemical pathways or nutritional requirements are being studied, defined media are preferred.

Culture media can be classified *nutritionally* as either minimal or all purpose (rich). A **minimal medium** supplies only the minimal nutritional requirements of an organism, no extras. For example, if the organism has the genetic information to produce all its own amino acids, none are added to the medium. If the organism normally requires two amino acids, those two are added, but no more. In each case, the medium is considered minimal, even though minimal medium can be different for different organisms. Many minimal media support only a relatively narrow range of organisms. An **all purpose** or **rich** medium is one which contains a wide variety of nutrients, far above the minimal requirements of the organism. A rich medium usually supports the growth of a wide range of organisms. Most organisms will grow faster and better on a rich medium than on a minimal medium because resources need not be diverted to the production of compounds such as amino acids, fatty acids, vitamins, and nucleic acid bases which the medium supplies.

Functional classification of culture media is done to identify the purpose for which the medium is used. Three classes are recognized: selective, differential, and selective-differential. A **selective** medium supports the growth of a desired organism, while inhibiting the growth of unwanted ones. Selectivity can be brought about in two ways: by adding a compound which inhibits the undesired organisms or by deleting a compound required by them. Inhibitory compounds that might be added include: dyes, such as crystal violet and many other triphenyl methane dyes, which inhibit the growth of Gram-positive organisms; antibiotics; sodium azide (NaN_3), which interferes with cytochrome function; and high levels of salt or sugar which affect osmotic pressure (refer back to the environmental parameters of growth, exercise 6). A **differential** medium is one which supports the growth of many different organisms, but *differentiates* between them. For example, the several media containing bromcresol purple used earlier (Exercise 8) support the growth of many different organisms, but can differentiate between them based on the ability to produce acid from sugar. A **selective-differential** medium is one which combines the benefits of both functional classes: it selects for a small group of organisms (by selecting against others), and at the same time differentiates between those organisms that *do* grow.

Although these media classifications are convenient, be aware of the overlap from one category to the next.

In this exercise, you will be studying the various classes of media. Much of this study requires that you refer to previous exercises. This exercise is also designed to give you more practice in streaking agar plates for pure culture isolation and to encourage class discussion.

Period A

Materials

1. One to two ml of each of the following cultures: *E. coli, K. pneumoniae,* and *S. epidermidis*
2. Three plates of eosin-methylene blue (EMB) agar and three plates of nutrient agar (NA)

Procedure

1. Label one plate of EMB and one plate of NA for each of the three organisms given. Now streak the appropriate organism onto each plate to obtain isolated colonies. Directions for preparing streak plates are in Student Supplement 1. Invert and incubate the plates at 37° for two days.
2. Before the next period, compile your observations (on Report Sheet 10) for the following:
   - Results from Exercise 9 of the growth of *E. coli* and *Lactobacillus* on minimal medium 1 (MM1) and all purpose Tween (APT) agar
   - Results from Exercise 8 of the growth of *E. coli* and *Micrococcus luteus* in the lactose Durham tubes

This period is intended to be one of discussion. Questions following the procedure should be answered by looking at the plates from the last period, compiled observations from earlier exercises, and the media list in Student Supplement 2. Final conclusions should be written on Report Sheet 10.

Materials

1. Plates streaked last period
2. Compiled observations from Exercises 8 and 9
3. Demonstration plates of *E. coli* and *Lactobacillus* on MM1 and APT agar and demonstration tubes of *E. coli* and *Micrococcus* in lactose Durham tubes for confirmation of earlier observations. These plates and tubes were prepared as described in the procedure sections of Exercises 8 and 9. There should be one set of demonstration plates and tubes for each 8 students.

Procedure

1. Look at the plates you streaked last time. Do you have isolated colonies? If not, ask the instructor to look at them and give some suggestions for improvement. It will be critical in future exercises that you be able to obtain isolated colonies. You should have at least 10 and preferably more like 30 well-isolated colonies.
2. Observe again the EMB and NA plates from last period and record the presence and appearance of growth on Report Sheet 10. All three of the organisms should have grown on the nutrient agar, but the *Staphylococcus* should not have grown on the EMB agar. Nutrient agar is considered nonselective. It is true that there are some organisms that will not grow on nutrient agar, but most of the common ones will. Also, while it is true that you might be able to differentiate the three organisms by colony morphology on nutrient agar, it is not considered to be a differential medium, since many organisms will not have markedly different colony morphologies on nutrient agar. EMB, on the other hand, is considered to be both selective and differential. The eosin and methylene blue dyes inhibit the growth of Gram-positive organisms, and the lactose and the dyes together allow for the differentiation of organisms based on their ability to ferment lactose. On EMB, *E. coli* should produce a dark purple colony with a green sheen on the upper surface of the colony. (Tilt the plate for best observation of the green sheen.) The green sheen is due to the formation of an eosin and methylene blue dye complex which occurs at low pH. In this medium, the low pH is caused by the production of large amounts of acids in the fermentation of lactose. *Klebsiella* produces a lighter colored colony, with a dark center. This colony is often called a fish-eye colony because of its appearance. *Klebsiella*, unlike *E. coli*, produces fewer acids and more neutral endproducts when fermenting lactose. Recall the results of the methyl red and Voges-Proskauer tests for these two organisms (Exercise 8). Also recall the

presence of a capsule around *Klebsiella* cells seen in Exercise 3, which gave the colony a slimy appearance. We thus see that the differential qualities of EMB agar extend beyond just the detection of lactose fermentation.

Questions

1. To study the chemical classification of media, look at the medium constituents of Minimal Medium 1 and nutrient agar in Student Supplement 2. *E. coli* can grow on both media. Which is the defined medium? What is it about the other that makes it undefined? Be sure to read the section on complex medium ingredients in Student Supplement 2.

2. To study the nutritional classification of media, look at your observations from Exercise 9. Can *E. coli* grow on both MM1 and APT agar? Does it grow better on one than the other? Remember, MM1 is the minimal medium for *E. coli*. What ingredients present in APT agar supply amino acids, vitamins, nucleic acid bases, and fatty acids?

3. If MM1 is the minimal medium for *E. coli*, why couldn't the auxotrophic mutant grow on it? What is the minimal medium for the auxotroph?

4. Is APT agar a minimal medium for *Lactobacillus*? How could you determine the minimal requirements for the growth of this organism?

5. To study the functional classification of media, look again at MM1. Is it selective?

6. The lactose Durham tubes used in Exercise 8 contain a differential medium. Using Student Supplement 2 as a reference, what are the differential agents in the medium? What is in the medium that prevents this medium from being classified as a selective one? (In other words, what is in the medium that allows a wide range of organisms to grow?) Is the small glass vial in the Durham tube considered a differential agent? Ah, this is a tricky question. The vial itself is really not part of the medium, but is certainly part of the overall differential system. The classification of the Durham-tube vial as a differential agent indicates some of the ambiguities involved in the classification of media. Does this mean all this classification is useless? No, it means that the classification of media is often open to discussion. However, it is certainly useful to be able to categorize a medium immediately as to its potential function.

7. Return to Student Supplement 2 and attempt to classify the media chemically, nutritionally, and functionally. As the course proceeds, you will find that close observation of the composition of media will often simplify the understanding of an exercise, whereas ignoring the composition of media may hinder understanding. Just as a chef must look at the ingredients in a recipe to predict whether the result will be chocolate cake or chicken cacciatore, a microbiologist must often study the medium constituents to predict and understand the results of an experiment.

8. Define the following types of media: defined; complex; rich; minimal; selective; differential; selective-differential.

9. Devise a medium that will allow the growth of *E. coli* but not *Bacillus subtilis* and that will differentiate between different strains of *E. coli* based on their ability to ferment galactose. (Help? Think about EMB agar and possible modifications.)

References

Guirard, B. M. and **E. E. Snell**. 1981. *Biochemical Factors in Growth*. In **Manual of Methods for General Bacteriology**. Gerhardt, P., editor. American Society for Microbiology. Washington, D.C. A fairly extensive discussion of the requirements for growth and the different types of media used. The first five or six pages of this chapter would be most appropriate for an introductory student.

Notes

Culture Media

Fill in the following charts to facilitate discussion.

From Exercise 9:
Score growth from − to 4+

Medium	*Escherichia coli* wild type	*Escherichia coli* auxotroph	*Lactobacillus bulgaricus*
Minimal medium 1			
All purpose Tween agar			

From Exercise 8:
Score as + or −

Result	*Escherichia coli*	*Micrococcus luteus*
Acid production in lactose Durham tube		
Gas production in lactose Durham tube		

From Exercise 10:
Score growth as + or − and describe the appearance of the colonies if growth occurred.

Medium	*Escherichia coli*	*Klebsiella pneumoniae*	*Staphylococcus epidermidis*
Eosin-methylene blue agar			
Nutrient agar			

(Continued on next page)

Fill in the following chart, classifying the media listed.

Medium	Defined or undefined (complex)	Minimal or all purpose (rich)	Nonselective, selective, differential, or selective-differential	Selective or differential agents
Minimal medium 1				
All purpose Tween agar				
Lactose Durham tube				
Eosin-methylene blue agar				
Nutrient agar				

Identification of Bacteria

The identification of bacteria is based on observations of cellular morphology, cultural morphology, physiology, and serological and bacteriophage typing. Characteristics of **cellular morphology** include cell shape, size, response to staining procedures such as the Gram stain, motility, flagellar arrangement, presence of capsules or spores, and cellular arrangement (formation of pairs, chains, filaments, and so forth). Characteristics of **cultural morphology** include colony shape, size, color and texture on a specified medium. Cellular and cultural morphology alone are not sufficient to identify a bacterial culture to genus and species; the **physiology** of the organism must also be studied. Physiological observations include oxygen requirements, substrate utilization, growth factor requirements, endproducts formed, ability to grow under various conditions of pH, water availability, and temperature, and ability to grow in the presence of certain inhibitory compounds. Taken together, morphological and physiological characteristics are usually sufficient to identify an organism to genus and species. Sometimes it is useful to identify a bacterial culture further to subspecies or strain, usually for tracing the sources of infectious diseases in populations. To do this, serological or bacteriophage typing of cultures may be done.

In this exercise, you will receive two unknown cultures from the list of cultures given in Table 11-1. You are to make cellular, cultural, and physiological observations on these unknowns, and then identify them using the information in the chart in Table 11-1. A complete discussion of the tests to be performed can be found in Student Supplement 1.

Table 11-1   Cultures used and their characteristics

Characteristic	Bacillus polymyxa	Bacillus subtilis	Escherichia coli	Klebsiella pneumoniae	Lactobacillus bulgaricus	Micrococcus luteus	Staphylococcus epidermidis
Gram reaction	+	+	−	−	+	+	+
Cellular morphology	rod	rod	rod	rod	rod	coccus	coccus
Cell arrangement	chains	chains	single	single	single	pairs	clusters
Spores	+	+	−	−	−	−	−
Catalase	+	+	+	+	−	+	−
Indole	−	−	+	−	−	−	−
Gelatinase[1]	+	+	−	−	−	+/−	−
Amylase[1]	+	+	−	−	−	−	−
O_2 requirements	facul-tative	aerobe	facul-tative	facul-tative	aero-tolerant anaerobe	aerobe	facul-tative
Glucose fermentation	A+G[2]	−	A+G	A+G	A	−	A
Lactose fermentation	A+G	−	A+G	A+G	A	−	A
Motility	+	+	+	−	−	−	−

[1]The gelatinase and amylase tests are often variable, depending on the medium used, length of incubation time, and other factors. For this reason, the cultures used in class should be tested before distribution to the students, and the results in the above table modified, if need be.

[2]A=acid; G=gas

Period A

Materials

1. Two ml of two different unknown cultures, grown overnight on a shaker
2. Two sterile pipettes
3. Two plates of nutrient agar and two plates of starch agar
4. Three tubes of each of the following media: gelatin medium, thiogly-collate agar, lactose Durham tube with bromcresol purple, glucose Durham tube with bromcresol purple, tryptone broth, and motility medium. The thioglycollate medium should be melted and the tubes held at 45–50° in a water bath until use.

Procedure

1. Record the code number of your unknown cultures.
2. Label a plate of nutrient agar for each unknown and streak each plate with the appropriate culture for isolated colonies.
3. Label a plate of starch agar for each unknown and inoculate the appropriate culture onto the plate in a single line down the center of the plate. *Do not* streak for isolated colonies.
4. Label one tube of each medium for each culture and retain one tube of

each medium as a sterile control. Inoculate all tubes except the tube containing motility medium with a drop of culture using a sterile pipette. It might be useful to practice pipetting with a nonsterile pipette and water before attempting to pipette the cultures. After inoculation, gently roll the tubes to disperse the inoculum. Cool the thioglycollate agar tubes in cold water as done in Exercise 6.

5. Inoculate the motility medium using the inoculating needle. Your inoculation line should be as straight as possible so you can determine movement away from the line after incubation as an indication of motility.

6. Prepare a smear of each unknown. Allow the smear to air dry, then heat-fix it. If time allows, perform the Gram stain and, if you still have time, observe the slide for cellular morphology. Record your observations on Report Sheet 11. If there is not enough time, complete these observations next period.

7. Prepare a wet mount of each unknown and observe it for motility. Record your observations on Report Sheet 11.

8. Incubate all plates and tubes at 30° *for 2 days only.**

Period B

In this period, you will be observing your unknown cultures for a variety of cultural and physiological characteristics. Refer to Student Supplement 1 and earlier exercises for specific procedures as needed. Be sure to fill in all the information required in the Report Sheet to make identification of your organism easier. When handling reagents, exercise care, as they may be harmful to the skin, and wash hands after use.

*Safety
Caution!*

Materials

1. Plates and tubes inoculated last period
2. Reagents for tests: Gram's iodine; Kovac's reagent; 5% hydrogen peroxide
3. Ice
4. Demonstrations of positive and negative results †

Procedure

1. If you have not already performed and observed the Gram reaction of your unknown culture, do so. Record observations in Report Sheet 11.
2. Perform the indole test by adding several drops of Kovac's reagent to the tube of tryptone broth. Do not mix. A red ring indicates a positive test.
3. Place the gelatin tubes in ice water. Allow them to sit for 15 minutes. These tubes will be observed and discussed in step 10 below.

4. Observe the thioglycollate tubes for oxygen requirements of your unknowns. (Refer to Exercise 6 for assistance.)

5. Observe the motility tubes. Look for turbidity (cloudiness) away from the line of the stab, indicating motility (use the uninoculated control tube to assess turbidity of the medium itself). The semisolid nature of the motility medium allows movement through the medium by motile organisms, but prevents nonmotile organisms from leaving the line of the stab. Do the results agree with the observations from the wet mount made last period?

6. Observe the lactose and glucose Durham tubes for production of acid and gas from each sugar. Record your observations.

7. Flood the starch plate with Gram's iodine, watching carefully for a clear area around the colony. Iodine reacts with starch to form a blue complex. Starch, a glucose polymer, can be degraded by microorganisms that produce the enzyme *amylase*. If the starch around a colony is degraded, there is nothing to react with the iodine, and a clear area is seen. Observation of the clear area must be done rather rapidly, as the color fades in a few minutes. Record your results. (A caution here: A positive test for starch is a negative test for the enzyme amylase; a negative test for starch is a positive test for amylase. Be sure you record the results properly.)

8. Observe the nutrient agar plate for colonial morphology, including colony shape, size and color. See Figure 11-1 for some helpful terminology for describing colony morphology. Record your results.

9. Test your cultures for catalase, an enzyme used by the cell to detoxify hydrogen peroxide formed during growth and contact with oxygen. The

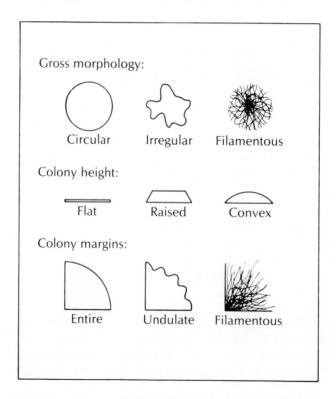

Figure 11-1 *Some helpful terminology for the description of colony morphology.*

enzyme breaks H_2O_2 into water plus oxygen, and its action can be visualized by dropping H_2O_2 on a colony and looking for bubbles. Perform this test on a few isolated colonies. Be sure to watch closely, as bubble formation may occur slowly. Slow formation of bubbles is also considered a positive reaction.

10. Observe the cultures for their ability to produce gelatinase. Gelatin is a high-molecular-weight polypeptide which is liquid at warm temperatures, but solid at 20° or less. Enzymes that break down gelatin are called *gelatinases* and are members of the protease group of enzymes. If gelatin has been degraded to its constituent amino acids or oligopeptides, the tube of gelatin will remain liquid even after chilling. To observe a negative test for gelatin degradation, tip the sterile control tube cooled on ice in step 3 and notice how the gelatin remains in place. Now slowly tip the cooled tubes of your unknowns and record the results.

11. Before the next period, compare your results with those listed in Table 11-1 and identify your two unknown organisms. Turn in these identifications along with a list of what *you* thought were the key tests for identification.

Questions

1. If you were given a slant of an unknown culture and asked to give a tentative identification of it in 15 minutes, what would you do? How reliable would your answer be if you knew the organism was from Table 11-1 and you identified it to be *Lactobacillus bulgaricus*? *E. coli*? *B. subtilis*?

2. In the above exercise you assayed either directly or indirectly for at least six enzymes. What are the enzymes and what are the tests?

3. Why is it important for the motility medium to have a lower agar concentration than conventional agar deeps?

4. Why did you test for motility in more than one way? Can you think of any circumstances in which the two tests might not agree? (Consider the effect of culture age and oxygen concentration on motility.)

5. In this exercise, you have tested for the production of the extracellular enzyme amylase. Why would a cell excrete this enzyme rather than transport the substrate into the cell for degradation there? (Consider the size of the starch molecule.) Which of the other enzymes studied are extracellular and which are intracellular?

6. Now that you have studied the physical properties of gelatin (i.e. liquid at temperatures above 20°) and its degradation by commonly encountered microorganisms, discuss the advantages of using agar as the solidifying agent in microbiological media.

References

Paik, G. 1980. *Reagents, Stains and Miscellaneous Test Procedures.* In **Manual of Clinical Microbiology**. 3rd edition. Lennette, E.H , A. Balows, W.J. Hausler, Jr. and J.P. Truant, editors. American Society of Microbiology. Washington, D.C. An easily used reference to the tests used above, with precautions indicated for clinical application.

*If the time between periods is longer than 2 days, arrangements should be made to have cultures refrigerated after 2 days until the next laboratory. This refrigeration step may make the catalase reaction occur more slowly, so the student should watch the colony especially carefully after the addition of the catalase reagent.

†Instructor: See Table 11-1 to determine which cultures to use as demonstrations for positive and negative results for each test.

Identification of Bacteria

Fill in the following chart.

Test or characteristic	Unknown #1	Unknown #2
Code number		
Gram reaction		
Cellular morphology		
Cell arrangement		
Spore production		

Motility
    in wet mount
    in motility medium		

Tryptophanase (indole test)		
Oxygen requirements		

Glucose fermentation:
    acid
    gas		

Lactose fermentation:
    acid
    gas		

Amylase		
Colony morphology on nutrient agar		
Catalase		
Gelatinase		
Identification		

Notes

Bacterial Genetics

As was noted in Exercise 1, not all the cells of a pure culture are genetically identical, even if they have arisen from successive divisions of a single parent cell. This genetic variability may arise as a result of mutations which occur in the DNA. Genetic changes can also occur as a result of gene transfer, either by transformation, conjugation, or transduction. Read about these topics in your text before performing this exercise.

The purpose of this exercise is to demonstrate some of the variations possible in a bacterial population and to introduce you to some laboratory procedures for the isolation of mutants and recombinants. Mutants and recombinants can be obtained by either direct or indirect methods. *Direct selection* methods are those which allow for immediate isolation of the mutant or recombinant because the parent cells cannot grow on the selection medium.* *Indirect selection* must be used whenever the parent can grow in all situations in which the mutant or recombinant can grow. For ease of manipulation, we will be selecting for genetic variants by direct selection, but the questions at the end of the exercise relate also to indirect selection methods.

In one part of your work in this exercise, you will be isolating mutants which are resistant to different antibiotics. Antibiotic resistance is a useful property for genetic selection, since the resistant mutants will grow but the nonmutant sensitive cells in the population will be unable to grow in the presence of the antibiotic. The antibiotic to be used is streptomycin, which binds to the small subunit of the bacterial ribosome and interferes with the translation of the mRNA into protein. Mutants can be streptomycin-resistant due to the formation of altered ribosomes, development of decreased permeability to the antibiotic, or even acquisition of the ability to detoxify the antibiotic by chemical modification. Cells with altered ribosomes have the highest level of resistance to the antibiotic.

For the second part of this exercise, you will study gene transfer by conjugation of cells in liquid or on a solid agar surface. In both methods, you will be looking for recombinant cells which have regained the ability to produce an amino acid that they could not produce as auxotrophic mutants.†

Period A

Materials

1. For isolation of mutants, one ml of a 20-fold concentrated culture of *S.*

epidermidis. The culture should be grown overnight at 37° on a shaker, and the cells can be concentrated by filtration or centrifugation.

2. For selection of mutants, one plate of nutrient agar containing 10 μg/ml streptomycin

3. One sterile pipette

4. For conjugation, two ml each of *E. coli Hfr* Thr$^+$, Str^S (i.e. capable of producing the amino acid threonine and sensitive to the antibiotic streptomycin) and *E. coli F$^-$* Thr$^-$, Str^R. These strains should be cultured in nutrient broth and incubated for 10–16 hours at 37° on a shaker.‡ *Note to the Instructor:* It is critical that these cultures be less than 18 hours old. The cultures should be dispensed into tubes immediately before class, and these tubes should be kept at 37° until used.

5. One empty sterile tube

6. For the conjugation experiment, two nutrient agar plates and 7 plates of a mineral salts plus glucose medium containing 50 μg/ml streptomycin**

7. For dilution, three tubes containing 9.0 ml sterile 10^{-3} *M* potassium phosphate buffer, pH 7

8. Eight sterile pipettes

9. A glass spreader and alcohol for flame sterilization

10. A 37° water bath to hold the tubes containing the conjugation mixture

Procedure

1. Conjugation in liquid medium: Carefully and slowly, pipette one ml of *E. coli Hfr* and *E. coli F$^-$* into the sterile empty tube. Mix gently, then leave *undisturbed* at 37° for at least 30 minutes during which gene transfer should occur.

   As controls, pipette 0.1 ml of *E. coli Hfr* onto one labelled plate of nutrient agar and one labelled plate of the mineral salts medium. Flame-sterilize the spreader and spread the inoculum over the plates. Repeat this with new plates and sterile spreader for the *E. coli F$^-$*. The nutrient agar plates will serve as positive controls for culture viability; the mineral salts plus glucose and streptomycin will serve as a control for *reversion* of the threonine auxotrophic marker of the *F$^-$* strain to prototrophy *in the absence of gene transfer*, as well as a control for mutation to streptomycin resistance by the *Hfr* strain.

2. Conjugation on agar: Using your loop, draw a single line of *E. coli Hfr* across a plate of the mineral salts plus glucose medium containing streptomycin. After resterilizing the loop, draw a single line of *E. coli F$^-$* at right angles to the first line (see Figure 12-1). Allow the inocula to dry before inverting the plate for incubation. Gene transfer should occur in the section of the plate where the two cultures are mixed. The sections where the cultures are not mixed will serve as controls for mutation, similar to those in step 1.

3. For isolation of antibiotic-resistant mutants: Pipette 0.1 ml of *S.epidermidis* onto a properly labelled plate of nutrient agar containing added streptomycin. Flame-sterilize the spreader and distribute the inoculum over the agar surface.

   Incubate these plates for 5 days at 37°.

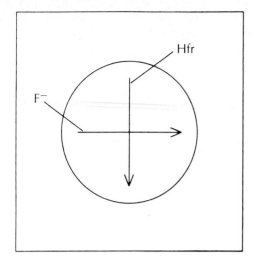

Figure 12-1 *Cross-streaking pattern for an* E. coli *F⁻* × Hfr *conjugation experiment. Each inoculated loop is drawn across the plate in the direction indicated by arrows.*

4. When at least 30 minutes have elapsed since mixing the two *E. coli* strains (step 1), shake the tube sharply for 30 seconds to disrupt the mating pairs.§ Immediately prepare three ten-fold serial dilutions of the mixture, using the sterile pipettes and the tubes of 9.0 ml sterile buffer. Remember to label the tubes, to pipette accurately, to mix each dilution well before removing a sample for the next dilution tube, and to change pipettes between each transfer. Figure 12-2 might help you with the dilution series. For help with dilution procedures, see Student Supplement 3.

After the dilutions are prepared, pipette 0.1 ml of each dilution or the undiluted mixture onto properly labelled plates of the mineral salts plus glucose medium containing streptomycin. For the platings, if you pipette from the *most* dilute tube to the *least* dilute tube, you need not change pipettes between steps. This should result in a 10^{-4} through 10^{-1} dilution series on the plates. Before the inoculum has soaked in, distribute it over the agar surface with a sterile spreader. If you spread from the 10^{-4} to the 10^{-1} plate, you need not sterilize the spreader between plates. (Why?) **Important!** If you cannot work rapidly, do the plate inoculation and spreading with a partner, so that the inoculum has no chance to soak in before spreading. If you work with a partner, one person should pipette while the other one spreads the inoculum.

Incubate all the plates in the conjugation experiment at 37° for 2 days.

Period B

In this period, you will be observing the plates you prepared last time in the conjugation experiment. Be sure to keep track of which plates were prepared to identify mutants of the parent cells and which were prepared to identify genetic recombinants.

Incubation of plates containing *S. epidermidis* cultures should be continued until Period C.

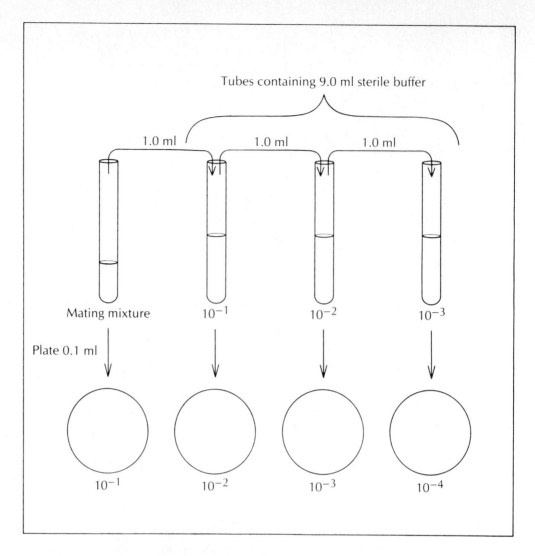

Figure 12-2 *Dilution and plating for enumeration of recombinants.*

Materials

Nutrient agar and mineral salts plates inoculated last period with *E. coli*

Procedure

1. Observe the nutrient agar plates of *E. coli Hfr* and *E. coli F⁻*. If the cultures were viable, you should see a haze of turbidity on the plates due to growth of thousands of colonies.

2. To check parent cultures for revertants to prototrophy or mutation to streptomycin resistance, observe the mineral salts plates on which the two cultures were separately plated. Count the number of revertants to prototrophy by the *F⁻* strain and the number of mutations to streptomycin resistance by the *Hfr* strain.

3. To check for recombinants, observe the plates prepared from the dilution series of the conjugation mixture. Count the number of recombinants

on the plate containing 30–300 colonies. Recombinants will be evident as colonies which can grow on the mineral salts medium because the cells are now capable of making threonine and are resistant to strep-tomycin.[¶] Calculate the number of recombinants per ml of the original conjugation mixture. In a successful conjugation experiment, the number of recombinants per ml of original mating mixture should be more than 100 times the number of mutants or revertants of the parents. Often, the number of recombinants is many times higher.

4. Observe the cross-streaked plate of *E. coli Hfr* and *E. coli F⁻* for presence of colonies on the part of the plate where the cultures were mixed. Are there any colonies in the sectors where the cultures were not mixed? This cross-streaking technique can be used as an easy method of testing a large number of cultures for conjugation.

Period C

In this period, you will be observing the plates of *S. epidermidis* for mutation to antibiotic resistance.

Materials

1. The plate prepared in Period A with *S. epidermidis*
2. One plate of each of the following media: Nutrient agar; nutrient agar containing 10 μg/ml streptomycin; and nutrient agar containing 100 μg/ml streptomycin

Procedure

Observe the *S. epidermidis* plate prepared last period. Can you see both large and small colony types? If so, count how many you see of each and record your observations on Report Sheet 12. What could account for the difference in colony size? (Hint: Consider the level of resistance that might be obtained if the cell has decreased permeability to the antibiotic compared to having an alteration in the site at which the antibiotic was active such that the antibiotic no longer affected the cell.) Divide each of the new plates into two halves, labelling one half for large colonies and the other for small colonies. Using your inoculating needle, pick some cells from one of the large colonies by just touching the needle to the top of the colony. (It is important that your needle be cool for this manipulation.) Now spot-inoculate each of the three media (nutrient agar with varying additions of streptomycin) with this needle. Transfer cells from another 2–10 large colonies onto the same half of each of these plates, then transfer 3–10 of the smaller colonies to the other half of the plates. Which colonies do you predict will grow on the high dosage antibiotic plates?

Incubate these plates at 37° for 2–5 days.

Period D

Materials

Plates inoculated last period

Procedure

On the *S. epidermidis* plates, observe the growth of transfers from colonies of different sizes. Record your results. Can you see a relationship between colony size on the low concentration of antibiotic and ability to grow on the high concentration? Were your predictions from Period C correct?

Questions

1. How would you differentiate between genetic change by mutation and as a result of gene transfer?
2. In this exercise, you selected for mutants *directly*. Based on any available information, devise a method for selecting mutants *indirectly*. For example, how would you isolate a streptomycin-sensitive mutant from the streptomycin-resistant cultures used in Period C?
3. Rifampicin, another antibiotic, binds to RNA polymerase and prevents initiation of transcription. (Cells sensitive to rifampicin will therefore be unable to synthesize RNA.) Would you expect a rifampicin-resistant mutant to be automatically resistant to streptomycin? Consider the site of action of the two antibiotics.
4. Did the antibiotic *cause* the mutation or simply *select* for mutants already present in the population?
5. Would it be possible to obtain a recombinant that is Thr$^-$ and Str^s? How would you select for this recombinant?

References

Carlton, B.C. and B.J. Brown. 1981. *Gene Mutation*. In **Manual of Methods for General Bacteriology**. Gerhardt, P., editor. American Society for Microbiology. Washington, D.C. This is a comprehensive treatment of procedures for mutagenesis and mutant selection, but it does not cover the specifics of the exercise done here.

Curtiss III, R. 1981. *Gene Transfer*. In **Manual of Methods for General Bacteriology**. Gerhardt, P. editor. American Society for Microbiology. Washington, D.C. A more advanced coverage of gene transfer than would be found in an introductory text.

Miller, J.H. 1972. **Experiments in Molecular Genetics**. Cold Spring Harbor Laboratory. Cold Spring Harbor, New York. An excellent laboratory manual covering many introductory and advanced topics in the molecular genetics of *E. coli*.

Bacterial Genetics

Mutation

Number of streptomycin-resistant colonies (Period B):

Number of streptomycin-resistant colonies of *Staphylococcus epidermidis* on nutrient agar containing 10 μg/ml streptomycin:

small _____

large _____

Levels of resistance to streptomycin (Period C):

Number of:	Streptomycin concentration in agar		
	0 μg/ml	10 μg/ml	100 μg/ml
Small colonies transferred from Period B which grew			
Small colonies transferred total			
Large colonies transferred from Period B which grew			
Large colonies transferred total			

Which type of colony had the highest resistance? _____

What is a likely cause of this resistance in the organism? Change in site of action? Permeability change? Production of inactivating enzymes?

(Continued on next page)

Gene transfer in *Escherichia coli*

Results of conjugation in cross-streaking experiment:

Shade in the area where growth occurred.

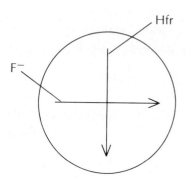

Results of conjugation in quantitative experiment:

E. coli F^- and *Hfr* viability tests on nutrient agar:

F^- (+ or −) _____

Hfr (+ or −) _____

Number of F^- Thr$^+$ revertants/ml of culture on mineral salts agar + glucose + streptomycin:

Number of *Hfr* Str^R mutants/ml of culture on mineral salts agar + glucose + streptomycin:

Number of recombinants (Thr$^+$, Str^R) on mineral salts agar + glucose + streptomycin after conjugation:

Dilution counted	Number of colonies	Recombinants/ml of original mating mixture

Is the number of recombinants/ml of original mating mixture at least 100 fold higher than the number of revertants or mutants?

Does this indicate a successful conjugation?

13

Regulation

The genetic information that an organism carries in its DNA, its **genotype**, ultimately determines the capabilities of that organism. The outward expression of this genotype is called the **phenotype** of the organism. The phenotype depends on the presence in the cell of proteins which are produced through the processes of transcription and translation of **genes** coding for these proteins. The proteins called enzymes catalyze the various reactions in the cell, including the production of various biochemical constituents. The synthesis of these enzymes is regulated by the cell through the processes of induction and repression. In addition, the activity of an existing enzyme can be regulated by mechanisms such as feedback inhibition. The regulation of enzyme synthesis and enzyme activity is affected by cellular events as well as by environmental factors. Therefore, the phenotype depends not only on the genotype but also on the environment in which the organism is growing. For example, the enzyme β-galactosidase, which breaks lactose into glucose and galactose, is produced only during growth in the presence of lactose. Sometimes, the phenotype is affected by subtler changes in the organism's environment, such as temperature or pH. This exercise is designed to demonstrate two phenotypic differences that occur as a result of environmental variations. The results of this experiment also should demonstrate the importance of *standardization* of cultural conditions when trying to characterize an organism.

Period A

Materials

1. One to two ml of cultures of *E. coli* and of *Pseudomonas fluorescens*
2. One plate of nutrient agar and one plate of Pseudomonas F agar
3. One tube containing 5 ml of nutrient broth plus 1% glucose and one tube containing 5 ml of nutrient broth plus 1% lactose

Procedure

1. Streak one plate of nutrient agar and one plate of Pseudomonas F agar with the *Pseudomonas* culture. Incubate the plates at 30° for 2 days.

2. Inoculate each of the nutrient broth tubes with a loopful of *E. coli* culture. Incubate these tubes at 37° for 2 days.

Period B

In this period, you will be observing the *Pseudomonas* plates for the production of fluorescein, a fluorescent compound used by the cell as an iron chelator. You will also be testing the *E. coli* cultures for the presence of the intracellular enzyme β-galactosidase. To test for the enzyme, the substrate ortho-nitrophenyl-β,D-galactoside (ONPG) will be used. If the enzyme β-galactosidase is present in the cell, the ONPG will be cleaved to form galactose and ortho-nitrophenol, a yellow compound. The appearance of the yellow ortho-nitrophenol is a positive test for the enzyme. To make the cells permeable to the ONPG substrate, they will be treated with toluene.

Materials

1. A source of ultraviolet light, such as a short-wavelength mineral light, set up in a darkened room
2. Two ml of a solution of 5% toluene in acetone
3. Two ml of a solution of $1.3 \times 10^{-3} M$ ortho-nitrophenyl-β,D-galactoside (ONPG) made up in $10^{-3} M$ potassium phosphate buffer, pH 7.6
4. Pipettes and pipette bulbs for pipetting the toluene and ONPG

Procedure

1. Pipette 0.5 ml of the toluene solution into each of the tubes containing nutrient broth plus glucose or plus lactose. Mix well and put the tubes aside for 10 minutes. The toluene will make the cells permeable to ONPG.

Safety Caution!

2. Observe the plates streaked with the *Pseudomonas* culture. Note the formation of a yellowish pigment on the Pseudomonas F agar plates. Observe this plate under an ultraviolet light in a darkened room. (**Caution!** Ultraviolet light irreversibly damages the retina of the eye after even short exposures from these lamps. **Do not** look directly into the UV lamp.) The pigment produced by this organism will absorb the UV radiation and emit light at a different wavelength, appearing fluorescent. The pigment is called fluorescein, and its formation is used as a major taxonomic characteristic in the differentiation of species in the genus *Pseudomonas*. Fluorescein belongs to a group of compounds called *iron-ophores*, which are organic molecules excreted by cells which are able to scavenge iron in the environment. In an aerobic environment, iron usually exists as an insoluble ferric hydroxide. Ironophores can solubilize the iron, making it available for transport into the cell, where it is used

104

in the making of cellular constituents such as cytochromes. Pseudomonas F medium has very low levels of iron, and it appears that this low environmental iron triggers the increased synthesis of the ironophore. The ability of an organism to synthesize an ironophore under conditions of low environmental iron should make it possible for this organism to compete successfully with other organisms in iron-poor environments. Nutrient agar has much higher levels of iron, and much lower amounts of fluorescein are made.

3. Ten minutes after the toluene has been added to the two nutrient broth cultures of *E. coli*, add 0.5 ml of the substrate ONPG to each tube. Mix well and observe for a bright yellow color indicative of the hydrolysis of the ONPG to ortho-nitrophenol and galactose by the enzyme β-galactosidase. The color should appear in 5 minutes or less. Record the results.

Questions

1. You have now studied the production of the enzyme β-galactosidase and of the biosynthetic product fluorescein. Recognize that the ability of an organism to produce enzymes such as β-galactosidase or those enzymes which produce fluorescein is determined by the cell's genotype, but that environmental factors can influence the *expression* of this genotype, through affecting either the synthesis or the activity of the proteins. Now, in your own words, differentiate between phenotype and genotype.

2. Describe the function of metal chelators excreted by microorganisms. Rely on your textbook for specific information. Do animals excrete metal chelators into their environment? (Think about the internal environments of the intestinal tract and blood stream.)

3. Why was ONPG hydrolyzed in the tube containing cells grown in the presence of lactose, but not in the tube containing cells grown in the absence of lactose?

4. What would the results of the β-galactosidase assay have been if the addition of toluene had been omitted?

References

Miller, J.H.. 1972. **Experiments in Molecular Genetics**. Cold Spring Harbor Laboratory. Cold Spring Harbor, New York. An excellent laboratory manual covering both genetic and regulatory exercises using *E. coli*.

Notes

Regulation

Production of fluorescein by *Pseudomonas fluorescens*

Is fluorescein produced on nutrient agar?

Is fluorescein produced on Pseudomonas F agar?

Explain the results:

Production of β-galactosidase by *Escherichia coli*

Medium	Color within 5 minutes after addition of ONPG?	Did the culture produce β-galactosidase?
Nutrient broth + glucose		
Nutrient broth + lactose		

Explain the results:

Notes

14

Action of Bacteriophage

A bacteriophage is a virus that infects a bacterial cell. Like all viruses, bacteriophages rely on the cellular machinery for energy production and for many of the functions of replication, transcription, and translation. The most dramatic type of virus infection is the lytic infection, in which the host cell is lysed when new virus particles are released. The lysis of many cells can easily be seen as a loss of turbidity in a broth culture or as a clearing called a **plaque** on an agar surface with otherwise confluent growth of bacteria. The plaque assay can be used to **titer** (enumerate) viruses in a suspension or to perform **bacteriophage typing** of various cultures of bacteria. Titering might be done in preparation for genetic work using transduction as the method of gene transfer or possibly in ecological studies of the virus itself. Bacteriophage typing is, as mentioned in Exercise 11, a tool in identification of bacterial strains and relies on the specificity of the host range of bacterial viruses.

In this exercise, you will titer a suspension of the virus T2 or T4 on its host *E. coli* B. You will also test other bacteria for their susceptibility to phage T4 or T2, to demonstrate both the bacteriophage typing of host cells and the host range of a virus. For further background to this exercise and the questions which follow, refer to your textbook.

Period A

In this period, you will be preparing lawns of bacterial cells using an agar overlay method. An agar overlay results in a more uniform lawn than that prepared by streaking and allows for easy visualization of plaques since the lawn is a very thin film over a clear bottom layer of agar. The bottom layer is necessary to prevent drying of the overlay during incubation.

Materials

1. One ml of a suspension of phage T2 or T4*
2. Two ml of an overnight culture of *E. coli* B and one ml of an overnight culture of each of the following: *E. coli* K, *Klebsiella pneumoniae*, and *Bacillus subtilis*.

3. Nine plates of bottom agar
4. Two tubes containing 9.9 ml sterile water and four tubes containing 9.0 ml sterile water for dilutions
5. Twelve sterile 1.0 ml pipettes
6. Nine tubes of top agar, approximately 3 ml per tube. The medium should be melted and the tubes kept at 45–50° in a water bath until ready for use.
7. If possible, a water bath or test tube hot block should be available for every 8 students to minimize the time the top agar tubes are out of the water bath. Alternatively, a temporary water bath can be made by using hot water from the tap. Be sure the water is above 45°.

Procedure

1. Label 5 plates for the dilution series for virus titering. You will be diluting the virus suspension to 10^{-8} and plating 0.1 ml of that to give 10^{-9} on the last plate. (The plates, then should be labelled 10^{-5} to 10^{-9}.)
   Label the other 4 plates for one of each of the cultures for the bacteriophage typing/virus host range tests.
2. Arrange the dilution tubes with the two tubes containing 9.9 ml sterile water to be used first. You will be making two 1/100 dilutions followed by four successive 1/10 dilutions (see Figure 14-1). Label the tubes.
3. *Changing pipettes between each tube*, dilute the virus suspension.
4. Obtain six tubes of top agar. Keep them at 45–50° in the available water bath or hot block. From this point on, you must *work rapidly* so that the agar overlay does not solidify until after it is poured into the plate.
   Into each tube of top agar, pipette 0.2 ml of *E. coli* B cells. Mix gently. Remove one tube from the water bath, add 0.1 ml of the 10^{-8} dilution of the virus suspension, mix gently and well, wipe any water off the outside of the tube, then pour the contents onto the plate marked 10^{-9}. Tip the plate quickly to spread the top agar over the entire surface. Put aside to cool. Using the same pipette, proceed to add, mix, and plate the 10^{-7} to 10^{-4} dilutions, each onto its appropriate plate. You should still have one tube of top agar left in your water bath that has been inoculated with host cells but no virus. Pour this tube into the plate marked for *E. coli* B alone and save for use in Step 5. Place all these plates out of the way in preparation for the rest of the experiment.
5. Obtain the last three tubes of top agar. These will be used for the experiment on bacteriophage typing. Inoculate each with one of the other three cultures, mix well, and pour into the appropriately labelled plate. Note that no virus has yet been added. Allow the overlay to solidify. After solidification, transfer a loopful of *undiluted* virus suspension onto the center of each agar surface including the plate of *E. coli* B with no virus prepared in Step 4. Allow the liquid to soak in before inverting the plates for incubation.
7. Incubate all plates at 37° for 2 days.

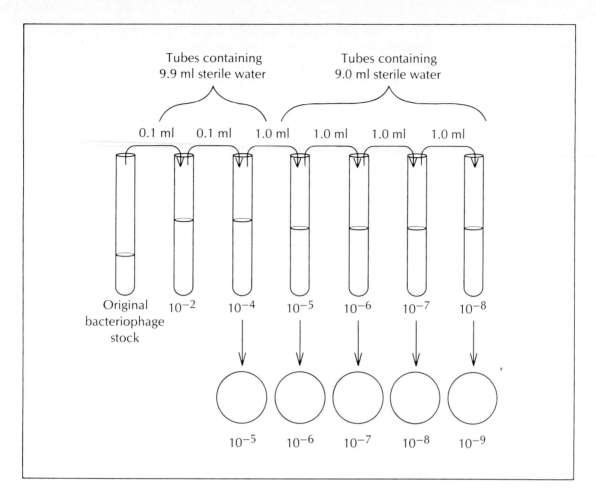

Figure 14-1 *Dilution and plating for bacteriophage titering. Add 0.1 ml of each dilution to tubes of top agar that have been inoculated with host cells. Pour onto plate of bottom agar.*

Period B

Materials

The plates from last period

Procedure

1. Observe the plates from the dilution series. Plaques should be clearly visible in one or more of the plates. If you have difficulty seeing them, hold the plates up to the light. Count the plaques in the plate containing 30–300 plaques and calculate the titer of the virus in the original suspension. If you need help, see Student Supplement 3. Is one of the virus-infected plates completely clear? This is called *confluent lysis*. A plate exhibiting confluent lysis is useful for preparing stocks of bacteriophages (see Appendix 2 on culture preparation).

2. Observe the plates from the bacteriophage typing/virus host range experiment. You should see an area of lysis on the *E. coli* B plate where the virus was placed. Do you see lysis on any of the other plates? What does that tell you about the specificity of infection of the bacteriophage used?

Questions

1. What is the purpose of the calcium in the top agar? (Hint: See Student Supplement 2).
2. Why is a thin agar overlay used instead of inoculation of the complete 15–25 ml of agar in the plate?
3. What do you think accounts for the specificity of infection of a virus for its host? For a few helpful hints on this question, consult your textbook to define the site of recognition and attachment of a virus to its host cell. Think also about the specificity of recognition of components in the replication, transcription and translation systems in a cell and the presence and activity of enzymes which degrade specific DNA molecules (the restriction enzymes).
4. If you continued to incubate the plates in Period B, the plaques would *not* increase in size.[†] Why not? Consider the requirements of the virus for infection and growth and for eventual liberation from the bacterial cell. The host cells must be actively metabolizing. In a 2-day-old lawn of bacteria, the cells are in stationary or death phase and biosynthetic reactions are very slow. Could the size of the plaques be limited by the diffusibility of the virus through the top agar? Look at the percentage of agar used in the top agar (Student Supplement 2). It is not high enough to cause significant diffusibility problems with the small virus particles.

References

Miller, J. H. 1972. **Experiments in Molecular Genetics**. Cold Spring Harbor Laboratory. Cold Spring Harbor, N.Y. An excellent laboratory manual which covers introductory and advanced techniques in virology.

*Any available virulent bacteriophage which is active against *E. coli* B can be used in this experiment. If the only available bacteriophage is virulent only against *E. coli* K rather than B, be sure to use strain K rather than B in the titering experiment in Period A, step 4.

†A predator of bacteria called *Bdellovibrio* was actually discovered on a bacteriophage plate by the observation of a plaque that increased in size after the first day of incubation. *Bdellovibrio* cells infect and lyse bacterial cells, but do not need the host cell's energy generation or replication, transcription and translation machinery. *Bdellovibrio* cells are small and very motile, so they can move through the top agar easily.

Action of Bacteriophage

Host specificity of bacteriophages

Circle which organism(s) was susceptible to the bacteriophage used:

Escherichia coli B

Escherichia coli K

Klebsiella pneumoniae

Bacillus subtilis

Explain the phenomenon and use of bacteriophage host specificity:

Bacteriophage titering

What was the number of infective virus particles/ml in the virus stock suspension?

How is titering of viruses different from enumerating bacterial cells?

Notes

114

Microbes in Nature

Up to this point in the course, you have worked mainly with pure cultures of microorganisms. These pure cultures had been obtained from their natural environment by either direct isolation (usually on selective media) or by prior **enrichment** to increase the population of the desired organism before isolation. Enrichment is usually necessary when the organism is a minor component of the microbial flora, or when there is no strong selection procedure available for the desired organism. In the following exercises, a number of enrichment cultures will be set up to isolate microbes from natural environments such as water, soil and plant surfaces. It is the purpose of the present exercise to illustrate the diversity of microorganisms present in natural environments.

This exercise is actually almost a repeat of the work done earlier in Exercise 2, with a slight difference: you. At this point in the course, your more sophisticated understanding of microbes and of their physiology and morphology will make observations of hay infusions and lake water much more meaningful. In addition, it is important to refresh your appreciation of the diversity of microorganisms found in natural samples. This exercise should be done at a leisurely pace to allow time for complete observations. Think about the significance of what is being seen. For even more fun (!), you are encouraged to bring in your own natural water sample. Water from a rain trough or gutter? Water from a farm pond? If the water is clear, it will probably be necessary to concentrate it by filtration. This is easily done and is explained in Appendix 2.

Materials

1. A hay infusion made up of a hay and lake or pond water mixture incubated for at least 5 days in the light
2. Lake or pond water samples

Procedure

For the procedure used for examining these samples, refer to the instructions given in Period B of Exercise 2. Spend some time looking for all the different morphological forms of bacteria. You might be able to see some larger gliding bacteria in the hay infusion or in a sample from polluted water. Some of these gliding organisms will belong to the genus *Sphaerotilus* or *Beggiatoa*. See your text for a full description of these organisms. Also look for spiral-shaped bacteria. Can you see any eucaryotic cells? Are they easier to distinguish from the procaryotes now? Do you see any evidence of interactions between cells? What do you think the advantages or disadvantages are for each of the interacting organisms?

Microbes in Nature

Record your observations of the hay infusion. Look for various morphological types of procaryotes (rods, cocci, spirilla, long gliding chains) and eucaryotes (algae, protozoa, fungi).

Notes

118

16

Photosynthetic Bacteria

Photosynthetic procaryotic cells are divided into two groups based on their mechanism of photosynthesis. Those producing oxygen (called oxygenic), carry out the so-called "plant type" of photosynthesis and belong to the group called **cyanobacteria**. These organisms evolve O_2 as a byproduct of photosynthesis when they use water as an electron donor for the conversion of CO_2 to organic carbon compounds. Those organisms which do not produce oxygen (called anoxygenic), carry out the so-called "bacterial type" of photosynthesis and are called by the more general name of **photosynthetic bacteria**. (This can be somewhat confusing, since the terms procaryote and bacterium are usually used interchangeably. In this case however, although the cyanobacteria are certainly procaryotic, they are grouped separately from the other photosynthetic procaryotes.) The photosynthetic bacteria do not evolve O_2 during photosynthesis and obtain the electrons to convert CO_2 to organic carbon compounds from sources other than water, such as H_2 or H_2S. When growing photosynthetically, these organisms are obligate anaerobes.

This exercise will involve the enrichment and isolation of some purple bacteria of the family *Rhodospirillaceae*. These organisms are widespread in nature, primarily because they are facultative phototrophs, able to grow phototrophically when light is available and oxygen is absent and heterotrophically when oxygen is present. They can be found in soil, lake mud, and various other easily accessible sources. Because of their tolerance to oxygen and their nutritional versatility, these organisms are especially suitable for enrichment in a class exercise. In order to enrich for them, a medium will be prepared which contains NH_3 as a nitrogen source and a nonfermentable organic compound as a carbon source. The primary energy source in the enrichment will be light.

Period A

Materials

1. One empty screw-capped bottle, approximately 30–60 ml capacity
2. A source of photosynthetic bacteria, such as lake mud, marsh mud, marsh water, or soil
3. A sufficient amount of photosynthetic mineral salts (PMS) medium (see Student Supplement 2) to fill the bottle. This need not be sterile for this

exercise, although for isolation of an organism from a specific environment, sterility would be required.

4. A source of light: either natural sunlight, lowered in intensity by filtering through a curtain or by using a north-facing window, or incandescent light bulbs (preferably not fluorescent lights).

Procedure

1. Place some inoculum in the bottle. If mud or soil is being used, put in approximately one gram of inoculum. If water is the source of inoculum, put in 2–3 ml of sample.
2. Add PMS medium to the bottle to fill it *almost* to the top. Leave a very small air bubble at the top to allow for expansion of the liquid when it heats up in the light. Cap the bottle and place it either on a windowsill or 15–30 cm from a 25 watt light bulb. Incubate until a pinkish color is seen in the medium, probably in about 10–14 days.

Period B

Materials

1. The enrichment culture prepared in Period A
2. One plate of photosynthetic mineral salts (PMS) agar
3. Some method of obtaining anaerobic incubation of the plates. Suggested methods are Gas-Pak jars (or other sealable containers) with Gas-Pak H_2/CO_2-generating envelopes or sealable containers which have an outlet for evacuation and provision for gassing with N_2/CO_2 (95/5 ratio). For this, a vacuum source and gas tank of N_2/CO_2 with appropriate regulator is needed.

Procedure

1. Make a wet mount of your enrichment. Look for motile rods (possibly *Rhodopseudomonas*), motile spirilla (possibly *Rhodospirillum*), and nonmotile barbell shaped cells (possibly *Rhodomicrobium*). Record your observations.
2. Look for evidence of phototaxis. To do this, watch one motile cell, then quickly cover and uncover the microscope light source. Did the cell reverse direction? Many of these organisms are *phototactic*, that is, they respond to the presence of light. Many motile photosynthetic bacteria make a reversal of direction when moving from light to dark (or even when moving from bright light to dimmer light). In your microscope preparation, when the light was eliminated, it mimicked the process of a cell moving into darkness. Note that the cell cannot "find" the light, but if its random movements should bring it into the light, the reversal

response effectively keeps it there.

3. Streak the plate for isolated colonies. Incubate the plate for 5–7 days *anaerobically* in the light.

Materials

The plate from the last period

Procedure

1. Observe the plate for colony morphology, including color. Are there any colorless colonies on the plate? What might these be? What is the cause of the different colors in the photosynthetic colonies? (See your textbook for a discussion of photosynthetic and accessory pigments.)
2. Pick a few colonies for wet mounts. Attempt to pick colonies of a few different colors or shapes. If you have any colonies that look like miniature red volcanoes, these might be *Rhodomicrobium*. Observe the wet mounts for cellular morphology. Record your results.

Questions

1. What is the purpose of the liquid enrichment in this exercise? Why wouldn't streaking a plate of photosynthetic mineral salts agar have been just as successful?
2. If these organisms are obligately anaerobic when growing photosynthetically, why didn't they all die when you streaked them out onto the plate in the (aerobic) laboratory room?
3. How would you set up an enrichment for cyanobacteria? How about an enrichment for a nitrogen-fixing cyanobacterium?
4. Why is it important that the organic carbon source be nonfermentable? (Consider the diversity of organisms that might grow in the presence of a fermentable carbon source.)
5. Identify the components of the enrichment system that made it selective for these photosynthetic bacteria. To answer this question, refer to your text for a discussion of other photosynthetic bacteria.
6. Considering the light absorption characteristics of the chlorophylls and accessory pigments in the *Rhodospirillaceae*, why would one type of lighting (incandescent) be more effective in an enrichment procedure than another (fluorescent)?
7. The genus *Rhodomicrobium* consists of photosynthetic bacteria which are *prosthecate* and reproduce by budding. Using your textbook or Bergey's

Manual as a reference, draw a cell group of *Rhodomicrobium* showing the prosthecae and buds.

8. Referring to exercise 5, explain why performing a viable plate count on a culture of *Rhodomicrobium* might underestimate the total number of cells present.

References

Buchanan, R. E. and **N. E. Gibbons**, editors. 1974. **Bergey's Manual of Determinative Bacteriology**. 8th edition. The Williams and Wilkins Company. Baltimore. This book contains the taxonomical classification of the procaryotes, excluding the cyanobacteria.

Photosynthetic Bacteria

Describe your observations of the enrichment and isolation below:

Wet mount of enrichment:

Did you observe phototaxis?

Wet mount of isolated colonies: Observe three different colonies and record your observations below:

	Morphology and color	Cell type
Colony 1		
Colony 2		
Colony 3		

(Continued on next page)

How was anaerobiosis achieved in the enrichment?

In the isolation?

Soil Microbiology

17

The soil contains many different microniches which support the growth of a large variety of microorganisms. These microniches vary in pH, oxygen conditions, water availability, and availability of nutrients. In addition, the microniches themselves change constantly as the weather conditions change (rain, drought, high temperature, low temperature) and as nutrient availability is affected by such diverse factors as leaf fall, plant growth, and insect activity. For this reason, most of the microorganisms found in the soil as normal inhabitants have some mechanism for long-term survival between periods favorable for growth. Some of these mechanisms are spore formation, encystment, production of protective capsules, and ability to go into prolonged metabolic dormancy. In addition to long-term survival mechanisms, many soil microorganisms have biochemical processes which allow them to compete effectively with other soil organisms for nutrients. For example, a microorganism may use a nutrient that many other organisms cannot, or it may produce a substance which is toxic to other organisms in the vicinity. Many soil procaryotes can use N_2 as their sole source of nitrogen. The conversion of N_2 to NH_3 by these organisms is called **nitrogen fixation** and is a process unique to the procaryotes. These nitrogen-fixing procaryotes can grow when many other organisms cannot grow because of a lack of combined nitrogen. Other microorganisms may use unusual organic substrates as carbon sources, thus allowing growth even when common substrates are absent. Some microorganisms produce toxic substances which may afford them competitive advantages over other organisms. One class of toxic substances produced by soil microorganisms are the antibiotics. An **antibiotic** is a compound produced by one microorganism that kills or inhibits the growth of another. Many of these antibiotics have been developed into therapeutic drugs to control diseases in animals.

In this exercise, you will be observing some of these survival and competition mechanisms, as well as some of the diversity of the organisms found in soil.

Period A

In this period, you will be setting up an enrichment for free-living (non-symbiotic) nitrogen-fixing bacteria of the genus *Azotobacter*, and preparing isolation plates for members of the genus *Streptomyces*. In addition, you will be plating a soil suspension to obtain growth of a variety of bacterial types, and you will be heating a suspension to obtain cultures of endospore-forming bacteria. If you wish, you may bring in your own soil samples.

Materials

1. A sample of fresh soil for the enrichment of *Azotobacter*. For best results, this soil sample should not be taken from a heavily fertilized soil, such as well-fertilized gardens or lawns.
2. A water suspension of fresh soil. To prepare, mix approximately one-tenth of a gram of soil with 10 ml of sterile water in a tube, shake vigorously, and allow the soil particles to settle.
3. A suspension of dry soil prepared as above. The dry soil can either be a soil sample that has been sitting around in the laboratory for a few weeks, or a sample from a dry area such as under a porch. Although soil from house plants can be used, the variety of organisms in this type of soil is fairly limited.
4. Two plates of nutrient agar and one plate of Streptomyces agar. The plates should have at least 25 ml of agar in them to prevent desiccation during the incubation period.
5. One bottle half-filled with nitrogen-free broth

Procedure

1. Streak a plate of nutrient agar with a loopful of the fresh soil suspension. Then put the tube containing the soil suspension in an 80° water bath for 15 minutes. (Be sure the water in the bath is high enough to cover the level of liquid in the tube.) This heat treatment kills vegetative cells, but leaves the heat-resistant endospores. Heating also triggers germination of spores and is referred to as *heat shocking*. Do not leave the suspension in the water bath for more than 15 minutes (Why?) and cool it rapidly after removing it from the water bath. Now streak the second (appropriately labelled) nutrient agar plate. Incubate these plates at 30° for 2 days.
2. Streak the plate of Streptomyces agar with the dried-soil suspension. Incubate this plate at 30° for 5–7 days.
3. Inoculate the nitrogen-free broth with fresh field soil. The inoculum should be approximately one gram per 100 ml of broth. Place the cap loosely on the bottle, thus allowing intake of N_2 from the outside and release of the gases formed in the bottle as organisms metabolize the sucrose in the medium. Incubate the bottle at room temperature for 5–7 days.

Period B

In this period, you will be observing the nutrient agar plates prepared last period for the variety of organisms present in soil (that will grow on this medium under these incubation conditions) and also for the presence of colonies of endospore-forming bacteria on the plates streaked after the heat treatment. The enrichment for *Azotobacter* and the isolation plates for *Streptomyces* must continue incubation for another few days before observations will be performed in Period C.

Materials

Nutrient agar plates inoculated last period

Procedure

1. Look at the morphology of the colonies that have developed on the nutrient agar plates inoculated last period. Compare especially the colonies on the plates prepared from the heat-shocked suspension with those from the untreated suspension. Prepare wet mounts of predominant types of colonies and note the cellular morphology. Can you observe any spores inside the cells? Prepare smears of a few of these colonies for staining in Period C.
2. Continue incubating these plates.

Period C

In this period, you will be studying spore-forming *Bacillus*, nitrogen-fixing *Azotobacter*, and filamentous *Streptomyces*. Plates will be prepared for the isolation of *Azotobacter* from the liquid enrichment cultures. In addition, you will be preparing antibiotic assay plates to test one of your own *Streptomyces* isolates for antibiotic production.

The antibiotic assay method you will be using involves streaking the (suspected) antibiotic-producer on an agar plate and allowing it to grow for 5–7 days. Since antibiotic production by *Streptomyces* occurs at about the same time as sporulation, the plates are incubated until the colonies acquire a chalky appearance indicative of spore production. At that time (Period E), you will be cross-streaking the plate with some other bacteria to assay for the presence of any antibiotic produced.

Materials

1. The plates and bottle prepared in Period A
2. Reagents for the spore stain
3. A boiling water bath for the spore stain
4. One plate of nitrogen-free agar
5. Three plates of tryptone-yeast extract agar. The plates should have at least 25 ml of agar in them to prevent desiccation during the incubation period.

6. One to two ml of a water suspension of *Streptomyces griseus* spores and 1-2 ml of a culture of *Bacillus polymyxa* for use as positive controls for antibiotic production in the antibiotic assay. These two bacteria are known producers of antibiotics.

Procedure

1. Prepare smears of some of the colonies from the heat-shocked suspension. Look for a colony which has a different appearance at the edge than in the center. This difference is, in part, due to the lysis of "mother" cells as endospores are liberated. This lysis can cause the colony to take on a collapsed appearance. Allow the smears to dry while performing steps 2 and 3.

2. Streak the plate of nitrogen-free agar with a loopful of material from the nitrogen-free broth. You are attempting to isolate *Azotobacter*, a strict aerobe, so your sample should come from the top of the liquid. Incubate the plate at room temperature for 2 days.

   Now prepare a wet mount from both the top and the bottom of the liquid (you will need a pipette, like a pasteur pipette, to obtain material from the bottom of the bottle). Compare the types of organisms present in the top and bottom of the bottle. How likely is it that all the cells you see fix nitrogen? (In other words, is it possible that some of the nitrogen fixed by the nitrogen-fixers is available to other organisms?) *Azotobacter* cells are football shaped (often in pairs) and larger than most of the bacterial cells you have previously observed. If you see *Azotobacter* cells, are they more prevalent at the top of the liquid? In the bottom of the liquid, look for cells with morphology characteristic of *Clostridium pasteurianum*, a nitrogen-fixing, endospore-forming anaerobe. Spores in this organism are found at the ends of the cells and swell the mother cells, giving them the appearance of spoons or tennis racquets. This organism may also play a role in the fixation of nitrogen in the anaerobic microniches in the soil.

3. Observe the Streptomyces agar plate for typical *Streptomyces* colonies. These colonies have a chalky appearance due to the aerial spores they form. Look for colonies of various colors. Color in *Streptomyces* colonies is shown in two different ways: the aerial mycelium and spores may be pigmented; or pigments excreted by filaments embedded in the agar may diffuse into the medium. Diffusible pigments are best seen by observing the colonies from the bottom of the plate.

   Choose one *well-isolated* colony to be used for the antibiotic assay. With a sterile loop, scrape the top of the colony for spores, then inoculate a tryptone-yeast extract plate with a single streak as shown in Figure 17-1. Inoculate the second tryptone-yeast extract plate with *Streptomyces griseus* and the third with *Bacillus polymyxa*. Incubate these plates at room temperature for 5–7 days.

   Observe the spore and spore-structure morphology of the *Streptomyces* colonies you have isolated. To do this, choose isolated colonies, but not the exact colony used for the antibiotic assay, as any spore-structures of this colony will have been destroyed by the loop. It is very difficult to prepare wet mounts of these organisms because of the compact nature

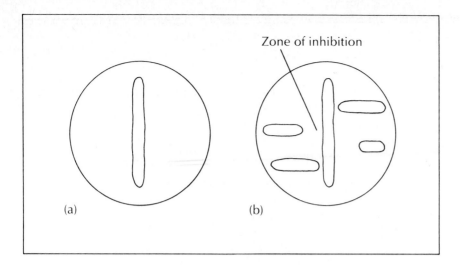

Figure 17-1 *Testing for antibiotic production. (a) Antibiotic producer is inoculated as a single streak and allowed to grow for 5–7 days; (b) Test cultures are cross-streaked. After two days of incubation, zones of inhibition can be seen.*

of their growth, but some idea of the morphology of spores and spore-structures can be seen by carefully pressing a dry coverslip down on a colony, then removing the coverslip from the colony and placing it down on a slide. Observe the slide with the high-dry lens (45×) to look for spores present in spirals, straight chains, or other configurations. The morphology of spores and spore-bearing structures is used in the characterization and identification of the many different species of *Streptomyces*.

4. Stain the smears from the 2 day and 7 day nutrient agar plates (see step 1, Period B and step 1 in this period) using the spore stain. The directions for this can be found in Student Supplement 1. Be sure to heat fix the slides first. Observe the smears using the 100× lens. Do you see green spores inside the red "mother" cells? Are there more free spores in the smears from the plates incubated for 7 days than in the smears from the plates incubated only 2 days?

Period D

Materials

The nitrogen-free plate from Period C

Procedure

Observe the nitrogen-free plate for typical colonies of *Azotobacter*. *Azotobacter* colonies are usually very slimy and brownish on this medium. Make a wet mount of slimy colonies and look for typical *Azotobacter* morphology.

Period E

In this period, you will only be cross-streaking your antibiotic assay plates with a few selected bacteria. Next period, you will be observing the area of intersection of the cross-streak with the center streak for inhibition of growth of these selected bacteria.

Materials

1. Tryptone-yeast extract plates inoculated last period
2. One to two ml of overnight cultures of: *E. coli, Micrococcus luteus, Pseudomonas fluorescens* and *Bacillus subtilis*

Procedure

1. Carefully cross-streak *E. coli* on the plate containing *B. polymyxa*, streaking inoculum from the *outside edge of the plate toward the center streak* to avoid picking up and carrying the antibiotic-producer into the streak of *E. coli*. (Why is this important?) Get the inoculum as close as possible to the center streak. Inoculate the other three cultures similarly.
2. Carefully cross-streak each of the four cultures onto the other two assay plates. Because of the possibility of contaminating the loop with the antibiotic-producer in each plate, *it is important* to flame the loop between each transfer. Incubate all plates at 30° for 2 days.

Period F

Materials

1. Plates prepared last period
2. A centimeter ruler

Procedure

Measure each cross-streaked culture for growth inhibition due to the production of antibiotics by the cultures streaked down the center. *S. griseus* and *B. polymyxa* are known producers of antibiotics, so some inhibition should be evident on those plates. Can you see any difference in inhibition between the two Gram-positive test organisms, *M. luteus* and *B. subtilis*, and the two Gram-negative organisms, *E. coli* and *P. fluorescens*?

Questions

1. What is the purpose of heat-shocking a soil suspension? Would *Streptomyces* spores survive this treatment?
2. Could you devise a medium and cultural conditions to isolate all the organisms from the soil at once? If not, why not?
3. Enzymes involved in nitrogen fixation are very oxygen sensitive, yet *Azotobacter* is a very aerobic organism. How can it fix nitrogen? Consider possibilities such as compartmentalization of enzyme functions and high respiratory rate.
4. Some *Klebsiella* species can fix nitrogen, but are unable to do it when the cells are growing aerobically. Referring to your answer to question number 3, what do you think is the reason for this?
5. What cellular macromolecules contain nitrogen? Based on this information, would you expect cells to require greater amounts of nitrogen than carbon for growth? Than sulfur? Iron?
6. Why was *dried* soil used as an inoculum source for *Streptomyces*? Consider the numbers of organisms in moist soil which could compete with the *Streptomyces* in the medium on which they were grown.

References

Krieg, N. R. 1981. *Enrichment and Isolation.* In **Manual of Methods for General Bacteriology**. Gerhardt, P., editor. American Society for Microbiology. Washington, D.C. This article contains useful information for the enrichment and isolation of a wide variety of bacteria.

Notes

Soil Microbiology

Isolation of endospore-forming bacteria from soil: *Bacillus*

Record your observations of *Bacillus* enrichments in the following chart:

Enrichment	Two-day-old culture		Seven-day-old culture	
	Colonies	Cells	Colonies	Cells
Not heated				
Heated				

Enrichment and isolation of *Azotobacter*

Observations of enrichment:

Sample from the top of the bottle

Sample from the bottom of the bottle

(Continued on next page)

Observations of the isolated colonies:

Colony morphology

Cellular morphology in a wet mount

Antibiotic production and the genus *Streptomyces*

Observation of isolations:

Observation of antibiotic production (growth inhibition of test organisms):

Culture	Test organism			
	Escherichia coli	*Pseudomonas fluorescens*	*Micrococcus luteus*	*Bacillus subtilis*
Streptomyces griseus				
Bacillus polymyxa				
Your isolate				

(Continued on next page)

Describe the morphological differences between the genus *Bacillus* and the genus *Streptomyces*.

General conclusions

What advantages can you suggest for the production of resting stages (for example, endospores by *Bacillus*, exospores by *Streptomyces*, cysts by *Azotobacter*) in the soil environment?

What advantages can you suggest for the production of antibiotics (for example, by *Streptomyces griseus* and *Bacillus polymyxa*) in the soil environment?

What advantages can you suggest for atmospheric nitrogen-fixation (for example, by *Azotobacter* and *Klebsiella*) in the soil environment?

Notes

136

Plant-associated Bacteria

Many bacteria are intimately associated with plants in both the terrestrial and aquatic environments. Some of these associations are true mutualistic **symbioses** where both the plant and bacterium benefit: for example, the association between nitrogen-fixing bacteria of the genus *Rhizobium* and leguminous plants. In some cases, the microorganism may obtain nutrients from the exudates of the plant in either the rhizosphere (area around the roots) or the phyllosphere (area around the leaves). The benefit to the plant from rhizosphere and phyllosphere bacteria is unclear, and in some cases, these bacteria are even plant pathogens. In this exercise, you will isolate bacteria from both the roots and leaves of plants. You will not need a preliminary enrichment for the desired organisms, as they are usually found in abundance on the plant materials. (The plant has already acted as a natural enrichment for the microorganisms.) The bacteria to be isolated from the roots are root-nodule bacteria of the genus *Rhizobium*. The leaf-associated bacteria isolated will probably be members of the genus *Pseudomonas*. You are encouraged to learn more about these organisms by referring to your text.

In this period, you will be attempting isolation of the nitrogen-fixing *Rhizobium* species from the root nodules of leguminous plants and fluorescent *Pseudomonas* species from the surface of leaves. The best sources of these organisms are, for the isolation of *Rhizobium*, *young* leguminous plants (e.g. peas, beans, alfalfa, soybean) which have obvious nodules, and, for the isolation of fluorescent pseudomonads, leaves which have not begun to senesce (deteriorate due to age). A suggested source of the plants is a vegetable garden. If that is not available, plants grown in a greenhouse may be suitable if they were planted in *nonsterile* soil. Plants growing in sterilized soil will have few bacteria associated with them.

Materials

1. Nodules and leaves as described above. If these are collected more than

137

a few hours before use, they should be kept moist by placing them in a plastic bag in a refrigerator until needed.

2. Forceps, glass rod, spreader and alcohol for flame sterilization
3. Two tubes, each containing two ml of sterile water, and one tube containing two ml of 70% alcohol. These tubes should be short enough for the forceps to reach the bottom.
4. One 250 ml flask containing 50 ml of sterile $10^{-3}\,M$ potassium phosphate buffer, pH 7, and a shaker platform to put the flask on
5. One sterile pipette
6. One plate of Rhizobium isolation agar and one plate of Pseudomonas F agar

Procedure

1. Using sterile forceps, put one or two leaves into the sterile phosphate buffer, being sure the buffer covers the leaves. Obviously, if you are isolating from corn leaves or other very large leaves, you will not use a whole leaf. Place the flask on a shaker for an hour while you work on the rest of this exercise.
2. Using sterile forceps, remove from the legume a root section containing one or two nodules. Rinse the root section in tap water to remove adhering soil, then place the root in the 70% alcohol. After 2 minutes, remove the root section, let the alcohol drain, then place the root in one of the tubes of sterile water. This procedure sterilizes the surface of the nodules.

Safety Caution!

3. Holding a slide with forceps, dip it into the alcohol and flame sterilize the slide. (Be careful that the burning alcohol does not run or drip onto hands or notebooks.) Place the slide on the lab bench and put on it a couple of loopfuls of sterile water. Using the sterile forceps, remove the nodules from the sterile water and place them in the water on the slide. Using the sterile (alcohol-flamed) glass rod, crush the nodules. Then streak a loopful of this material onto the Rhizobium isolation agar.
4. After the plant leaves have been shaking for an hour, remove the flask from the shaker. Remove 0.1 ml of the liquid and spread it on the Pseudomonas F agar plate.
5. Incubate both plates for 2–5 days at room temperature. (Why not 30 or 37°? Think about the normal environmental temperatures at which the plants grow.)

Period B

Materials

1. Plates prepared last period
2. An ultraviolet light source as used in exercise 13, in a darkened room

Procedure

1. Hold the Pseudomonas F agar plate prepared last period under the UV light and look for the presence of colonies of fluorescent *Pseudomonas* species. Avoid looking directly into the UV source as it can cause retinal damage. Some of these fluorescent pseudomonads may be plant pathogens such as *P. syringae*; others may be saprophytes such as *P. fluorescens*.

2. Observe the plate of Rhizobium isolation agar for the presence of typical slimy, clear *Rhizobium* colonies. The plate should contain almost solely *Rhizobium* colonies, unless the surface sterilization was not effective. Make a smear of one or two of these colonies and perform a Gram stain. *Rhizobium* cells should be Gram-negative rods, similar in size to *Pseudomonas*.

Safety Caution!

Questions

1. Why is combined nitrogen added to the Rhizobium isolation agar if rhizobia are nitrogen-fixers? (Does their proper title of "symbiotic nitrogen-fixers" help any?)

2. Since there are fluorescent pseudomonads (the generic term for organisms like *Pseudomonas*) on the surface of leaves, discuss why the surfaces of leaves don't fluoresce when exposed to a UV light source. Two explanations are: 1) the plant supplies sufficient iron so that the organism does not produce much fluorescein (remember from Exercise 13 that fluorescein is an ironophore); and 2) the number of cells present on the leaves is not sufficient to produce enough fluorescein to be seen. How would you determine which explanation is likely to be correct?

References

Krieg, N. R. 1981. *Enrichment and Isolation*. In **Manual of Methods for General Bacteriology**. Gerhardt, P., editor. American Society for Microbiology. Washington, D.C. This article contains useful information for the enrichment and isolation of a wide variety of bacteria.

Notes

Plant-associated Bacteria

Isolation of fluorescent pseudomonads

Record observations of Pseudomonas F agar platings of leaf washes:

What tests could be done to confirm that these fluorescent colonies belonged to the genus *Pseudomonas*?

Isolation of *Rhizobium*
Record observations of the plating of crushed root nodules onto Rhizobium isolation agar:

(Continued on next page)

What was the Gram reaction and cellular morphology of *Rhizobium*?

Ecologically, how do *Rhizobium* and *Azotobacter* differ?

Food Microbiology

19

Food microbiology deals with the study of both destructive processes (food spoilage) and constructive processes (food fermentations). Food spoilage includes the obvious spoilage which renders a food undesirable (it smells, looks, or tastes bad) and the sometimes less obvious spoilage resulting from the growth of food poisoning organisms, rendering the food dangerous to eat. In this exercise, you will determine the bacterial content of some foods, both fresh and not so fresh. To do this in a reasonable period of time, it is suggested that each person or pair of students analyze one type of food, with the results of the different types collated and discussed in a subsequent laboratory period. You will also make yogurt, a product of milk fermentation, and monitor the total bacterial population, the population of lactic acid bacteria, and the pH of the milk during the process of yogurt formation. Yogurt is a fermented milk product in which **lactic acid bacteria** produce acids which cause the milk to curdle (clot) and other substances which give yogurt its characteristic odor and flavor.

Period A

Materials

1. Food samples such as fresh raw hamburger, raw hamburger that has been refrigerated for a week, spices such as black pepper, cooked chicken (refrigerated one day), chicken salad prepared from the same cooked chicken (and also refrigerated one day): one sample per student
2. One bottle containing 100 ml sterile water for each student, used to prepare a slurry of the food to be tested
3. Blenders for blending "chunky" foods. If the blender container must be re-used during the period, it should be washed thoroughly with soapy water and disinfected with 70% ethanol for 5–10 minutes, followed by a sterile water rinse.
4. Three tubes containing 9.9 ml sterile water and seven tubes containing 9.0 ml sterile water, for dilutions
5. Eleven sterile 1.0 ml pipettes for dilutions and one sterile *wide-tip* 1.0 ml pipette for initial pipetting of the food slurry
6. A glass spreader and alcohol to sterilize it
7. Five plates of plate count agar (PCA)
8. 110 ml of pasteurized milk containing 2% fat

143

9. One clean jar
10. Two grams of powdered milk
11. Yogurt cultures: either 2 ml of *Lactobacillus bulgaricus* and 2 ml of *Streptococcus thermophilus* cultures grown overnight in sterile skim milk; or one tablespoon of *unpasteurized* yogurt. (Caution. Some commercially available yogurts are pasteurized before sale. Read the label.)
12. A method of determining pH, either a pH meter or, more simply, pH paper that covers the range of 4-7
13. Four plates of APT agar and four plates of APT plus sodium azide (NaN_3) agar
14. One or more 45° water baths with sufficient capacity to accommodate all the yogurt jars

Procedure

1. Add 100 ml of milk to the clean glass jar. Add two grams (about a teaspoonful) of powdered milk to supply extra growth factors for the bacteria. Mix well. Cover the jar *loosely* and steam the milk for 30–40 minutes. This steaming step can be done as a class group. The milk is usually steamed before inoculating to denature the milk proteins, thus making them more easily curdled.

   Use the other 10 ml of milk to determine its pH, as instructed. Record your results.

   After the milk has steamed for 30–40 minutes, the jars should be moved to the 45° water bath for quick cooling. They can remain there until you are ready to continue with yogurt production.
2. Prepare a slurry of the food sample to be assayed by mixing one gram of the food with the 100 ml sterile water. If the food sample is chunky, the sterile water should be poured into a sterile blender container, the food added, and the mixture blended at high speed until a uniform slurry is formed (approximately one minute). Remove 0.1 ml of the slurry and place it in 9.9 ml of sterile water. Pipetting may be difficult if the slurry contains chunks large enough to plug the pipette tip, so the special wide-tip pipettes should be used here. Prepare four subsequent 1/10 dilutions (see Student Supplement 3) of this dilution using the tubes containing 9.0 ml sterile water, being careful to change pipettes between each dilution and to mix each dilution well before proceeding to the next.
3. Pipette 0.1 ml of each of the last five dilutions onto plates of plate count agar marked with the final dilutions. (What are the final dilutions? By this time in the course, you should be able to set up your own dilution chart so you know what the final dilution will be in the plate. Need help? Refer to Student Supplement 3.) Spread the inoculum over the surface of the agar with a sterile glass spreader.
4. Returning to yogurt production, inoculate the 45° milk with the **starter culture** that has been described in item 11 of the Materials section. Mix well to distribute the inoculum. The major cause of failure in yogurt production is usually incomplete mixing. For the viable count procedure, remove 0.1 ml of the newly inoculated milk and put it into a tube containing 9.9 ml sterile water. This will result in a 1/100 dilution which will be used in step 5 below.

Replace the inoculated milk in the water bath for incubation for 3–4 hours, or until the milk has solidified.*

5. Returning to the viable count procedure for the yogurt preparation, mix the 1/100 dilution made in step 4 and proceed to make another 1/100 and three 1/10 dilutions using appropriately sized dilution tubes (remember to use a fresh sterile pipette for each dilution).

   Plate 0.1 ml of the last four dilutions onto labelled plates of APT agar and APT plus NaN_3 agar. Be sure to spread the inoculum evenly.

6. Incubate the PCA plates at 30° and the APT and APT + NaN_3 plates at 37° for 2 days.

Period B

In this period, you will be counting the colonies on the plates prepared in the previous period and preparing new dilution plates of the yogurt. Calculate your results as number of colony-forming units per gram of food or per ml of milk. As you observe these plates for microbial enumeration, remember that only those viable cells which can grow on the medium and under the cultural conditions used will form colonies. The absence of bacterial colonies on a particular medium *does not* necessarily guarantee that the food was sterile or even that the food is safe to eat, since there are a variety of food poisoning organisms, such as anaerobes and viruses, which could not be monitored in this way.

Materials

1. Plates prepared last period
2. Yogurt prepared last period
3. One bottle containing 100 ml sterile water, one tube containing 9.9 ml sterile water, and four tubes containing 9.0 ml sterile water for dilutions
4. Six sterile 1.0 ml pipettes and one sterile 10 ml pipette
5. A glass spreader and alcohol for sterilization
6. Four plates of APT agar and four plates of APT + NaN_3 agar
7. A method for determining the pH of the yogurt
8. A spoon for eating the fresh yogurt

Procedure

1. Count the colonies on the plate count agar plates prepared last time. Take note of the variety of organisms present. If time allows, perform a Gram stain of a representative number (at least two) of the predominant colony types. Calculate the total number of colony-forming units per gram of food.
2. Count the colonies on the APT agar as an indication of the total number of bacteria present in the steamed, cooled, inoculated milk used to make

yogurt. Count the colonies on the APT + NaN$_3$ agar as an indication of the total number of lactic acid bacteria in the milk. The NaN$_3$ makes this medium selective for lactic acid bacteria. For a discussion of the action of NaN$_3$, refer to your textbook. Calculate the number of colony-forming units per ml of milk.

3. Remove one ml of the yogurt with the 10 ml pipette and pipette it into the bottle containing 100 ml sterile water. If the milk is very solid, stir the yogurt to break up the curd enough for pipetting. Shake the dilution bottle well, then proceed to dilute the mixture further with one 1/100 and four 1/10 dilutions. Plate out 0.1 ml of the last four dilutions onto both APT and APT + NaN$_3$ agar plates. Incubate these at 37° for 2 days.

4. Determine the pH of the yogurt as directed by your instructor.

5. *Do not eat the yogurt in the laboratory!* If you wish to eat the yogurt, take it out of the laboratory.

Period C

Materials

1. Plates prepared in Period B
2. A compilation of plate counts obtained in Period B from the whole class

Procedure

1. Count the colonies on both the APT and APT + NaN$_3$ plates prepared in period B. Calculate the total number of bacteria and the total number of lactic acid bacteria per ml of yogurt. Compare this to your results from fresh milk. Are the numbers from yogurt higher? Lower? The numbers would be expected to be higher, because the cells grew and divided during the incubation period in the 45° water bath. However, the numbers may actually be lower due to the death of cells during refrigeration, probably resulting from the presence of acid, a toxic microbial end product.

2. Review the bacterial counts of the various foods sampled. Is there a significant increase in the microbial number after hamburger has been refrigerated for one week? Was this expected, in view of the fact that low temperature merely slows microbial growth? What about the difference between the counts in the chicken compared to the chicken salad?[†] What about the spices? High counts? Why?

Questions

1. What are the lactic acid bacteria and what is their significance in food?

146

How can they be selectively cultured?

2. What are the reactions occurring in milk during its conversion to yogurt?

3. How would you determine if a food is bacteriologically safe to eat (i.e. contains no food poisoning organisms or their products)?

4. Does a high bacterial count in a food necessarily indicate that it is unsafe to eat?

References

Peterson, C. S. 1979. **Microbiology of Food Fermentations**. 2nd edition. Avi Publishing Co. Westport, Connecticut. This book contains complete and easy to read discussions of many different food fermentations.

*The instructor should remove the jars for the whole class and refrigerate them until the next period. If the inoculum is active, the jars will be ready to refrigerate (i.e. the milk will have curdled) in less than 6 hours.

†If properly done, the counts should be lower in the chicken salad. Why should the chicken salad have lower counts, in view of the fact that meat and potato salads are supposed to be the culprits of many food poisoning cases? Actually, the low pH of mayonnaise used in many of these salads is inhibitory to bacterial growth. The real problem with the salads is apparently with the difference in handling of the salads at a hot summer picnic compared to the handling of bratwurst, hotdogs, hamburgers, and other foods. Most people are apparently careful to refrigerate the meats, but not the salads. In addition, meats are usually cooked well before being eaten. Also, even though the mayonnaise may be inhibitory, it is not uniformly distributed throughout the salad, and areas not in contact with the dressing, such as in the meat itself, may support microbial growth. Thus, food poisoning is still possible from mayonnaise salads.

Notes

148

Food Microbiology

Bacteria in a food sample

Food sample	Number of bacteria/gram	Colonial morphology of predominant types	Cellular morphology of predominant types
1.			
2.			
3.			
4.			
5.			
6.			

Were any of the above food samples unsafe to eat?

What would need to be done to determine this?

Yogurt production

pH:

before inoculation _____
after inoculation _____

(Continued on next page)

Viable count (bacteria/ml of milk or yogurt):

Medium	At inoculation	At conclusion of experiment
All purpose Tween agar		
All purpose Tween agar + NaN$_3$		

What is the ratio of lactic acid bacteria/ml to total bacteria/ml at the the end of the experiment compared to at inoculation?

Water Analysis

20

The bacteriological examination of water is performed routinely by water utilities and many governmental agencies to ensure a safe supply of water for drinking, bathing, swimming and other domestic and industrial uses. This examination is intended to identify water sources which have been contaminated with potential disease-causing organisms. Such contamination generally occurs either directly by human or animal feces or indirectly through improperly treated sewage or improperly functioning sewage treatment systems. Since human fecal pathogens vary in kind (bacteria, protozoa, viruses) and in number, it would be impossible to test each water sample for each pathogen. Instead, for drinking water, an **indicator organism** is used which only suggests the possible presence of pathogens by indicating the presence of fecal material or surface water contamination. To be useful as an indicator, an organism must conform to the following criteria: 1)it must be present in feces from both healthy and diseased individuals; 2)it must not be found in the uncontaminated environment; 3)it must persist in the water longer than the pathogens but not so long as to indicate an historical event; and 4)it must be safe and easy to isolate and identify. In the United States, the standard indicator is a group of organisms called the coliform group.

The **coliform** group is comprised of Gram-negative, nonspore-forming, aerobic to facultatively anaerobic rods, which ferment lactose to acid and gas at 35° in 48 hours. The major organisms in the coliform group are *E. coli*, *Enterobacter aerogenes*, and *Klebsiella pneumoniae*. The latter two organisms, however, can also multiply in the environment. Therefore, the presence of coliforms in a water sample is only presumptive evidence of fecal contamination or surface water contamination. A subgroup of the coliforms, the **fecal coliforms**, consists of Gram-negative, nonspore-forming, facultatively anaerobic rods, which ferment lactose to acid and gas at 44.5° in 48 hours. The only true fecal coliform is *E. coli*, which is found only in fecal material from warm-blooded animals. The presence of this organism in a water supply is evidence of recent fecal contamination and is sufficient to order the water supply closed until tests no longer detect *E. coli*. The total coliform count is used to assess the safety of drinking water, whereas untreated surface waters are evaluated with the fecal coliform count.

In this exercise, you will be testing water samples for the presence of both fecal and nonfecal coliforms. You are encouraged to bring in water samples of your own from wells, ponds, rivers, and lakes.

Period A

This period is designed to distribute equipment and information for obtaining your own water sample. If all the water samples for this exercise are supplied by the instructor, omit this period and start directly at Period B.

In order to obtain a sample for bacteriological assay, the sample must be taken aseptically. For this reason, sterile, empty bottles are provided, as well as instructions for obtaining a sample aseptically.

Materials

Two sterile screw-capped bottles which can hold at least 100 ml of water

Procedure

1. *Do not open the bottles until taking the samples! Do not take the water samples more than 24 hours before analysis in Period B.* When sampling tap water, use only cold water for sampling as it will be more indicative of the bacterial contamination of the water supply than will water which has been heated. (Water in a water heater is usually hot enough to kill a significant portion of the bacterial populations for which you are testing.)
2. To obtain a surface water sample, remove the cap of the bottle, retaining the cap in your hand. Now dip the lip of the bottle into the water (lake, river, puddle, and so forth) and allow the bottle to fill. Do not allow any water to wash over your hands and get in the bottle. Fill the bottle completely, replace the cap and screw down tightly.
3. To obtain a tap water sample, open the faucet and let it run for 5 minutes at a rapid rate. This washes away any bacteria on the faucet outlet so your sample will be of the water supply, not of the contamination of the faucet outlet by hands or whatever.

   Now, slow the water flow. Carefully open the sample bottle, retaining the cap in your hand as you would a tube cap. Do not let the lip of the bottle touch the faucet or anything else. Slowly and carefully, fill the bottle to the top. Replace the cap and screw down tightly.
4. Keep the two samples refrigerated until analysis in Period B.

Period B

In this period, two series of tubes will be set up containing dilutions of different water samples. For each dilution, six tubes will be inoculated: three for the coliform test and three for the fecal coliform test. One series of tubes will be incubated at 35° and the other at 44.5°.* In the next period, these tubes will be used to calculate the most probable number of coliforms and of fecal coliforms in the sample of water.

Materials

1. Fifty ml of each of two different water samples: one of tap water or well water and one of some surface water such as pond or river water
2. Twelve tubes of double-strength and 48 tubes of single-strength lactose lauryl tryptose broth Durham tubes
3. Six tubes containing 9 ml of sterile water for dilutions
4. Two 5 ml and seven 1 ml sterile pipettes

Procedure

1. Using three 9 ml dilution tubes for each water sample, prepare 3 successive 1/10 dilutions of each sample. Be sure to change pipettes between each dilution. (Help? See Student Supplement 3.) You now will have an undiluted sample and three dilutions: a 10^{-1} dilution, a 10^{-2} dilution, and a 10^{-3} dilution.
2. Using the 5 ml pipette, place 5 ml of the undiluted tap water sample into six of the tubes containing double-strength lactose lauryl tryptose broth. This 5 ml dilutes the double-strength broth to single strength and also allows a sample size of greater than one ml of water without the need to concentrate the sample by filtration. Mix by carefully rolling the tube in your hands. Do not mix so vigorously that air can enter the small vial inside the tube.
3. Using a new pipette for the undiluted sample and for each dilution of the tap water sample, pipette one ml of the undiluted sample into six tubes of single-strength lactose lauryl tryptose broth. Repeat for each of the dilutions. Mix carefully by rolling the tubes.
4. Separate the tubes so that three tubes of each dilution of the tap water sample are incubated at 35° and three of each are incubated at 44.5° Continue incubation for 2 days.
5. Repeat steps 2 through 4 for the surface water sample and its dilutions.

Period C

In this period, you will be observing the tubes for the presumptive presence of coliforms and fecal coliforms based on the production of gas in the broth. Although the presence of sodium lauryl sulfate in the broth makes it selective for enteric bacteria, further confirmatory tests would have to be done for a positive identification of coliforms. These further tests include pure culture isolation on selective media such as eosin methylene blue or MacConkey's agar by streaking samples from any lactose tubes which have acid and gas.

Resulting colonies would then be stained with the Gram stain to show the presence of Gram-negative, nonspore-forming rods. Time does not permit these confirmatory tests.

Materials

Tubes from last period

Procedure

1. Observe every tube for the presence of gas, indicated by the presence of a bubble in the small inverted tube. Record the results.
2. Using the Most Probable Number (MPN) table (Table 20-1), calculate the most probable number of coliforms and fecal coliforms in each water sample. This table represents a statistical evaluation of the probability of finding a given number of organisms in a sample for any given series of results. To illustrate the use of the MPN table, some idealized results are given in Table 20-2. In example A, the number of tubes showing gas was 3 out of the 3 inoculated with the undiluted sample ($10^0 = 1$ which indicates an undiluted sample), 2 out of 3 inoculated with the

Table 20-1   Three-tube most probable number (MPN) table

Combination of positives	MPN per inoculum of the middle dilution in the combination of positives which were used
0—0—0	<0.03
0—1—0	0.03
0—2—0	0.062
1—0—0	0.036
1—0—1	0.072
1—1—0	0.073
1—1—1	0.11
1—2—0	0.11
2—0—0	0.091
2—0—1	0.14
2—1—0	0.15
2—1—1	0.20
2—2—0	0.21
2—2—1	0.28
2—3—0	0.29
3—0—0	0.23
3—0—1	0.39
3—0—2	0.64
3—1—0	0.43
3—1—1	0.75
3—1—2	1.20
3—2—0	0.93
3—2—1	1.50
3—2—2	2.10
3—3—0	2.40
3—3—1	4.60
3—3—2	11.00
3—3—3	>24.00

Table 20-2 Examples of procedure for calculating MPN values

Example	Dilution								Selected combination of positives to use with the MPN table	MPN of coliforms/ml
	10^0		10^{-1}		10^{-2}		10^{-3}			
	Number positive	Number negative	Number positive	Number negative	Number positive	Number negative	Number positive	Number negative		
A	3	0	2	1	1	2	0	3	2–1–0	0.15×10^2 = 15
B	3	0	3	0	2	1	1	2	3–2–1	1.5×10^2 = 15
C	2	1	1	2	0	3	0	3	2–1–0	0.15×10^1 = 1.5

10^{-1} dilution, 1 out of 3 inoculated with the 10^{-2} dilution, and 0 out of 3 inoculated with the 10^{-3} dilution. Listing the positive results *in order*, gives 3–2–1–0. Choosing the three numbers in this series which just reach 0 (2–1–0), use the Table 20-1 to determine the most probable number of coliforms in the original sample. Note that 2–1–0 gives an MPN of 0.15 coliforms per inoculum (which was 1 ml for this dilution) of the middle dilution in that series, or 0.15 coliforms per ml $\times 1/10^{-2}$ which equals 15 coliforms/ml of the water sample. Try the other examples given in Table 20-2.

Record the results of your own water samples.

Questions

1. What is the sanitary significance of fecal contamination of surface waters? (Consider the likely sources of fecal contamination, including domestic animals and septic tank back-ups, and their involvement as sources of human pathogens which are shed in the feces.)

2. Could the MPN test be used to determine the number of lactic acid bacteria present in milk? Devise a liquid medium you could use; what would be considered a positive indication for their presence? (Hint: Remember, the lactics ferment glucose and are easily selected for by the addition of an agent that inhibits respiring organisms. See Exercise 19.)

References

Franson, M. A., editor. 1980. **Standard Methods for the Examination of Waste and Wastewater**. 15th edition. American Public Health Association. Washington, D.C. This book contains all the standard methods used by water utilities and governmental agencies for both microbiological and chemical examination of water.

Oblinger, J. L. and **J. A. Krueger**. 1975. *Understanding and teaching the most probable number technique*. J. Milk Food Technol. 38:540–545. Contains a detailed coverage of the MPN technique.

*These incubations are best carried out in water baths rather than air incubators as temperature can be controlled more accurately. If not enough water baths are available for the incubation, the tubes should be brought to the proper temperature in a water bath before being put in the air incubator.

Notes

Water Analysis

Coliforms

Number of positive (gas produced) tubes out of 3 inoculated and incubated at 35°

Water source	5 ml undiluted	1 ml undiluted	1 ml of 10^{-1} dilution	1 ml of 10^{-2} dilution	1 ml of 10^{-3} dilution
Tap water					
Surface water					

Most probable number of coliforms in tap water _____

Most probable number of coliforms in surface water _____

Fecal coliforms

Number of positive (gas produced) tubes out of 3 inoculated and incubated at 44.5°

Water source	5 ml undiluted	1 ml undiluted	1 ml of 10^{-1} dilution	1 ml of 10^{-2} dilution	1 ml of 10^{-3} dilution
Tap water					
Surface water					

Most probable number of fecal coliforms in tap water _____

Most probable number of fecal coliforms in surface water _____

Would either water be considered unsafe for drinking?

Notes

21

Microbes of the Body: The Cocci

In previous exercises, bacteria were studied that are found in the "natural" environment. The bodies of humans, animals, insects, molluscs, and other representatives of the 'animal kingdom' are also natural habitats for microorganisms. Many microorganisms exist in balanced symbiosis with host organisms. That is, the microorganisms grow on the surface of the host (surfaces include, of course, the "internal" surfaces of the gastrointestinal, genitourinary, and respiratory tracts) without causing measurable damage to the host. In some cases, the microorganisms actually benefit the host by preventing access to tissues by pathogenic microorganisms, by production of some useful compounds, or by affecting the host in some other way. There are also those microorganisms which can cause severe damage to the host. These organisms are termed **pathogenic** because of their ability to cause disease. **Virulence** is a measure of the *degree* of pathogenicity of the microbe on that specific host. A pathogen may be more virulent on one host than another. Because we have avoided the use of virulent pathogens in this course, the organisms studied in this exercise will be microbes which generally live continuously on or in the human body. Reactions expected from virulent pathogens will be demonstrated with less virulent or *avirulent* (not virulent) varieties of the organism, if possible. *You are cautioned, however, to be careful even with these less virulent varieties,* as many of them are **opportunistic** pathogens which *could* cause disease if the conditions are right. Warnings will be given throughout the exercise. Remember to use good aseptic technique (do not spatter inoculum when flaming the loop; wipe the bench top before and after working; wash hands before leaving the classroom).

One of the major morphological groups which can be found associated with animal bodies are the cocci. Some of the most important cocci are the Gram-positive cocci belonging to the genera *Staphylococcus* and *Streptococcus* and the Gram-negative cocci belonging to the genus *Neisseria*. One species of the genus *Staphylococcus* is *S. epidermidis*. You have already worked with this organism, which is commonly found on the skin of humans, and which has rarely been associated with disease. *S. aureus*, on the other hand, is associated with a wide variety of disease states, including impetigo, toxic shock syndrome, and food poisoning. This organism is found in the normal flora of the nasopharynx of about 40% of the human population. This shows that disease does not necessarily result simply from the presence of the disease-causing agent, but is also affected by other factors such as the physiological state of the host and the virulence of the particular strain of organism present. The genus *Streptococcus* also contains both pathogenic and

Safety Caution!

159

nonpathogenic members. A good example of a nonpathogenic streptococcus is *S. thermophilus* which you used to produce yogurt in Exercise 19. This organism, like many other lactic acid bacteria, is associated with the normal flora of both animals and plants, and generally causes no adverse effects. Other streptococci may be very pathogenic, however; for example, *S. pyogenes* and *S. pneumoniae* cause a wide range of diseases including pneumonia, scarlet fever, and acute rheumatic fever. Likewise, in the Gram-negative cocci of the genus *Neisseria* both nonpathogenic and pathogenic members occur: for example, *N. meningitidis* and *N. gonorrhoeae* are the causative agents of meningitis and gonorrhea, respectively. Another genus of Gram-negative cocci, *Branhamella*, is also commonly isolated from the body.

Differentiating between these groups of cocci is easy: *Neisseria* and *Branhamella* are oxidase-positive, Gram-negative, and rather kidney-bean-shaped cells often found in pairs; *Staphylococcus* are Gram-positive, catalase-positive cocci usually found singly or in clusters, and *Streptococcus* are Gram-positive, catalase-negative cocci often found in chains. The Gram-negative genera *Neisseria* and *Branhamella* can be distinguished because *Branhamella* species are unable to produce acids from carbohydrates while *Neisseria* species can.

Differentiating between the species within a genus is more complex, but obviously necessary for proper diagnosis of a disease. Since representatives of each genus are found in normal, healthy humans, diagnosis of a disease requires isolation of the disease-causing organism and subsequent identification to species, not just to the genus. Table 21-1 lists some major distinguishing tests used to differentiate species within each genus. In this exercise, you will perform some of these tests with pure cultures and will hopefully begin to appreciate the amount of work a clinical diagnostician must expend to identify an organism with certainty.

Exercise 21 is divided into two sections: 1)isolation of cocci from the human body; and 2)positive identification of the cocci. Since there is a chance of isolating a virulent pathogen from your body, you will not actually continue working with the cultures you have isolated, but will use laboratory cultures instead.

Table 21-1   Some tests used in the differentiation of species within the various genera of cocci

Genus	Tests
Staphylococcus	Coagulase
	Hemolysis
	Mannitol fermentation
	Production of heat-resistant extracellular nucleases
Streptococcus	Growth at 10°, 45°, or pH 9.6
	Hemolysis
	Production of levan or dextran from sucrose
	Sensitivity to optochin
Neisseria	Acid from carbohydrates
	Nitrate or nitrite reduction
	Production of levan or dextran from sucrose
Branhamella	Nitrate or nitrite reduction
	Cellular lipid analysis

In this period, the isolation of cocci from the body will be carried out. In addition, known organisms will be inoculated into a variety of test media, as required for the identification procedure.*

Materials

1. Overnight cultures of the following organisms: *Streptococcus salivarius*, *Streptococcus mitis*, *Streptococcus faecalis*, *Streptococcus faecalis* subsp. *zymogenes*, *Staphylococcus aureus*, *Staphylococcus epidermidis*, *Neisseria sicca*, and *Branhamella catarrhalis*

2. For the study of known cultures:

    - Eight blood agar plates
    - Four sterile optochin disks
    - Four plates of sucrose gelatin agar
    - Two plates of Vogel-Johnson agar
    - Two tubes of heart infusion broth
    - Eight sterile cotton swabs for streaking the plates
    - Two tubes each of Neisseria fermentation medium with glucose and Neisseria fermentation medium with sucrose

3. For the isolation of cocci from the body:

    - One blood agar plate
    - One plate of sucrose-gelatin agar plus NaN_3
    - One Vogel-Johnson agar plate
    - Two sterile cotton swabs

4. Forceps and alcohol for sterilization

Procedure

1. Using the sterile swabs, streak a blood agar plate with each of the cultures to obtain isolated colonies. After streaking, use an inoculating needle to make a cut 1 cm long in the agar on the *right* side of the most heavily inoculated area (the area of the first phase of streaking) of each blood agar plate. This is done to allow visualization of hemolysis by some of the oxygen sensitive **hemolysins** which cannot lyse blood cells on the highly aerobic surface of the plate.
    Next, using sterile forceps, place an optochin disk on the *left* side of the most heavily inoculated streaking phase of each *Streptococcus* plate. Do not use optochin disks with the other cultures.
    Incubate the blood agar plates in a candle jar† at 37° for 2 days. The candle jar ensures a higher CO_2 atmosphere, which is necessary for the growth of some of these organisms.

3. Streak a plate of sucrose gelatin agar with each of the *Streptococcus*

cultures. Incubate these at 37° for 2 days.

4. Streak a plate of Vogel-Johnson agar and inoculate a tube of heart infusion broth with each of the *Staphylococcus* cultures. Incubate at 37° for 2 days.

5. Inoculate the Neisseria fermentation tubes with the *Neisseria* and *Branhamella* cultures. Incubate for 2 days at 37°.

6. Prepare a Gram stain of each of the known cultures.

7. *Carefully* swab your throat with a sterile swab. Attempt to swab the back of the throat where the tonsils are (or were). This will likely result in a slight "gag response." It may be helpful to have the instructor or a fellow student obtain the sample for you, as one of them can see your throat easier. Alternatively, use a mirror to see your throat. Being careful not to let the swab dry out, streak a plate of blood agar to obtain isolated colonies. Incubate this plate with the other blood agar plates in the candle jar.

Safety Caution!

8. *Carefully* swab your nasal tract with a sterile cotton swab. You are attempting to obtain a sample from the nasal tract itself, so do not swab the opening of the nostril, but rather from the area at or immediately behind the constricture you can see if you look into your nostril. **Caution!** Do not swab any further back, as you risk hitting your olfactory bulbs, which can be easily injured (the pain will be considerable if you hit them!). Remember you are sampling in a gentle and *controlled* manner. Now streak a plate of Vogel-Johnson agar with this swab. Incubate the plate at 37° for 2 days.

9. Using the sterilized loop, scrape some of the whitish matter from between your teeth (preferably the back teeth) and streak a plate of sucrose-gelatin-NaN$_3$ agar with it. Incubate the plate at 37° for 2 days.

10. Discard all swabs properly so that they can be properly decontaminated before they are sent to the trash.

Safety Caution!

11. Wash your hands before leaving the laboratory.

Period B

In this period, you will be observing colonies of the known cocci and comparing them with your isolates obtained from the nose, throat and teeth. You will also be given unknown cultures to identify.

The media used in this exercise have been formulated to show characteristics of the different cocci. Blood agar is used to detect hemolysins, which lyse red blood cells. Hemolysis will be evident as a clearing of the agar due to complete hemolysis (termed *β*-**hemolysis**) of the red blood cells, or by a greenish coloration of the agar due to incomplete hemolysis (termed *α*-**hemolysis**). As mentioned earlier, some hemolysins are oxygen-sensitive, so be sure to check the cut area on the right side of the area of confluent growth for hemolysis reactions.

Sucrose-gelatin agar contains a high concentration of sucrose, a disaccharide, which can be split by some organisms, with one of the resulting monosaccharides (glucose or fructose) being polymerized (to dextran or levan, respectively), giving the colony a slimy appearance. The high concentration of gelatin apparently allows the slime to retain water and thus be more visible.

Vogel-Johnson agar contains lithium chloride, glycine and potassium tellurite, which make the medium selective for the staphylococci. The medium also contains mannitol, a fermentable carbohydrate, and phenol red as a pH indicator. Some species of staphylococci ferment mannitol and this property can be detected on Vogel-Johnson agar.

For safety, do not open the plates that were inoculated with material from your body.

Safety Caution!

Materials

1. For the coagulase test: one ml of rabbit plasma, two empty tubes, three sterile 1.0 ml pipettes and a 37° water bath
2. For the catalase test: 5% H_2O_2
3. Oxidase reagent
4. Uninoculated tubes of Neisseria fermentation medium plus glucose and Neisseria fermentation medium plus sucrose should be available in the laboratory to use as comparisons with the inoculated tubes.
5. Two unknown slant cultures plus enough media and optochin disks for identification‡

Procedure

1. Prepare smears of the unknown cultures for staining in Step 10.
2. Set up the coagulase test for each of the two *Staphylococcus* cultures. Label one empty tube for each culture, and add 0.5 ml of rabbit plasma to each. Then add 0.5 ml of the 2-day-old liquid culture of *S. aureus* to one labelled tube, and add 0.5 ml of the *S. epidermidis* culture to the other tube. Do not pipette these cultures by mouth! Use a pipette bulb. Incubate these tubes in the 37° waterbath and observe periodically for coagulation of the plasma during the laboratory period. (This test may require more time than is available during the laboratory period, and the instructor should make necessary arrangements for students who cannot complete this procedure during the period.)
3. Observe the blood agar plates of *all* the organisms being studied. Note the colony morphology, color and size. Record the results. Look for the presence of *hemolysis*. Rate it as α- or β-hemolysis. Remember, some of these hemolysins are oxygen-sensitive, so be sure to check the location where the agar was sliced before identifying the hemolysis type. Perform a slide-catalase test on each of the cultures by removing some of the cells with a loop and suspending them in a drop of 5% H_2O_2. Observe for bubbles. The catalase test cannot be performed directly on the plate because blood cells are catalase-positive!
4. Look for zones of inhibition around the optochin disks on the plates containing *Streptococcus* cultures. Optochin, which is not a therapeutically useful antibiotic, is a diagnostic tool used for the separation of the

Safety Caution!

alpha-hemolytic pneumococci (which are pathogens) from the other alpha-hemolytic streptococci. The pneumococci are optochin-sensitive, and thus cannot grow near the disk.

5. Observe the plates of sucrose-gelatin agar. Look for slime formation by the *S. salivarius* culture. Compare this with the plate streaked from your teeth. Do you see any slime producers?

6. Observe the plates of Vogel-Johnson agar streaked with the pure cultures of staphylococci. Note the distinctive black colonies formed by *S. aureus*. What is the basis of this reaction? (See Student Supplement 2.) Also note the yellow halo surrounding the colony. Compare these plates with the plate prepared from your own nasal sample.

7. Check the Neisseria fermentation tubes for acid production, using uninoculated controls to see color differences.

8. Test your known cultures for oxidase, an enzyme in the cytochrome system of many respiring organisms. The test is performed by dropping oxidase reagent onto a colony and looking for a dark pink or purple color that should appear in less than a minute. Cultures that have this enzyme, cytochrome oxidase, will exhibit the color change. This test is used as a taxonomic tool, not as a diagnostic test for respiration. Refer to your text for further discussion of this subject. Record your results.

*Safety
Caution!*

9. Observe the blood agar plates from your throat sample. *Without opening any plates*, compare the colony types with those of the known cultures. Can you make a tentative identification? What would be your next step in a clinical situation?

10. Prepare Gram stains and perform catalase tests on your unknown cultures. Based on the Gram reaction and the catalase reaction, obtain the appropriate media from your lab instructor. For example, if the unknown is Gram-positive, you will not need Neisseria fermentation medium; if the unknown is Gram-negative, you will. Additionally, a negative result of the catalase test will indicate the need for testing the culture for optochin sensitivity. (Why?) All unknowns should be tested on blood agar for hemolysis.

11. Check the coagulase test before the end of the period.

Period C

In this period, you should complete the identification of your unknown cultures.

Materials

1. For the coagulase test: rabbit plasma, empty tubes, sterile pipettes, a 37° water bath
2. 5% H_2O_2
3. Oxidase reagent

Procedure

Make observations and perform tests as required to complete identification of your unknowns.

Questions

1. What is hemolysis? What causes it? How can it be determined?
2. Why was the blood agar plate cut?
3. What is the function of the NaN_3 in the sucrose-gelatin agar used for the isolation of slime-producing streptococci from the mouth?
4. How could you differentiate between the following pairs of organisms:
   - *Neisseria* and *Streptococcus*
   - *Staphylococcus* and *Streptococcus*
   - *Neisseria* and *Branhamella*
   - *Staphylococcus* and *Neisseria*
5. Describe the order of events in the isolation and identification of cocci from samples obtained from the human body.
6. What is the role of the slime-producing streptococci in tooth decay?
7. How is it possible for healthy individuals to be carriers of pathogenic organisms?

References

Lennette, E.H., A. Balows, W.J. Hausler and J.P. Truant, editors. 1980. **Manual of Clinical Microbiology**. 3rd edition. American Society for Microbiology. Washington, D.C. This manual contains chapters on the isolation and identification of clinically important cocci. It is very useful for both the theoretical aspects and the practical aspects of these procedures.

*To save time and media, the inoculation of known organisms can be done by groups of four students. However, each student should observe all the procedures and test results. The cultures should be divided among the students: the two *Staphylococcus* cultures for one student, the *Neisseria* and *Branhamella* cultures for the second student, the four *Streptococcus* cultures for the other two.

†A candle jar is simply a large, sealable container in which plates can be incubated. After placing the plates in the container, a lighted candle is inserted, and the jar sealed. The candle will burn, using O_2 and producing CO_2, until the level of O_2 no longer supports combustion.

‡Since each student will perform a Gram stain and a catalase test on each unknown before beginning inoculation, not all media will be needed for each unknown. Students should select only those media which their preliminary data indicate they require. For each *Staphylococcus* culture, one tube of brain-heart infusion broth, one plate of Vogel-Johnson agar, and one plate of blood agar should be available. For each *Streptococcus* culture, one plate of blood agar, one plate of sucrose-gelatin agar, and one optochin disk are needed. For each *Neisseria* or *Branhamella* culture, one plate of blood agar and one tube of each of the Neisseria fermentation media should be supplied.

Notes

Name _____

Date _____

Report Sheet 21

Microbes of the Body: The Cocci

Characterization of pure cultures

Test or character*	*Staphylococcus epidermidis*	*Staphylococcus aureus*	*Streptococcus mitis*	*Streptococcus faecalis*	*Streptococcus faecalis* subsp. *zymogenes*	*Streptococcus salivarius*	*Neisseria sicca*	*Branhamella catarrhalis*	Unknown #1	Unknown #2
Gram reaction										
Cell morphology										
Catalase										
Oxidase										
Optochin sensitivity	✕						✕			
Hemolysis										
Colonies on Vogel-Johnson agar			✕	✕	✕	✕	✕			
Colonies on sucrose-gelatin agar	✕							✕		
Coagulase			✕	✕	✕	✕	✕	✕		
Acid from glucose	✕	✕	✕	✕	✕	✕	✕			
Acid from sucrose	✕	✕	✕	✕	✕	✕	✕			

*If test is not done, indicate by "ND."

Unknown #1 is _____   Unknown #2 is _____

(Continued on next page)

167

Observations of isolation of cocci from the body

Describe the colonies seen in each of the following isolations:

Throat swab on blood agar:

Nasal swab on Vogel-Johnson agar:

Gum/teeth swab on sucrose-gelatin agar + NaN$_3$:

If you had β-hemolytic colonies on the blood agar plate, are you sick?

If you had black colonies, surrounded by yellow halos, on the Vogel-Johnson agar, should you be worried?

Microbes of the Body:
The Enterics

The term **enterics** refers to organisms belonging to the family *Enterobacteriaceae*. This is a family of Gram-negative, oxidase-negative, facultatively anaerobic rods. Several groups are included in this family: genera such as *Escherichia, Klebsiella* and *Proteus*, members of which are generally found as normal inhabitants of the human body but which may also cause disease under certain circumstances; *Salmonella, Shigella* and *Yersinia*, usually associated with a disease state in animals; and *Erwinia*, often associated with a disease state in plants. In addition, various members of the genera *Erwinia, Klebsiella, Enterobacter, Hafnia, Serratia*, and *Proteus* are common inhabitants of water or soil. Because of their importance in human disease, the *Enterobacteriaceae* have been studied extensively. These studies have resulted in the availability of a wide diversity of diagnostic procedures involving many different media and tests which can be used to differentiate among the various organisms. The applicability of each of these individual tests to any given system varies according to the organism and the purpose of isolation.

In clinical situations, rapid identification of a suspected pathogen is needed. For this reason, clinical identification of the enterics usually starts with isolating the organism on a strongly selective and possibly differentiating medium, followed by serological identification of the isolate.* Serological identification is done using a slide agglutination test in which different commercially available antisera are individually mixed with cells from the isolate. If the isolate has antigenic determinants in common with those against which the antiserum was prepared, the cells will agglutinate (clump). This gives the clinician a rapid identification of the isolate, after which appropriate medical treatment can be started. The clinical microbiologist will then continue to confirm the identification with biochemical tests and will also perform an antibiotic sensitivity test on the isolate to confirm the appropriateness of the medical treatment. Such an early serological identification may give an epidemiologist a head start on searching out the source of the infectious agent if an epidemic threatens.

In this exercise, you will be studying the enterics from a clinical point of view. As mentioned at the beginning of this manual, the use of potential human pathogens will be minimized. For this reason, although the complete study of *Salmonella* and *Shigella* would give a good view of the clinical microbiologist's task, these organisms will be used only for the demonstration of serological identification of microorganisms, since serological identification can be performed with killed organisms. We will not carry out the serological identification of the other organisms in this exercise because the antisera are not as available as those against the serious pathogens.

22

Period A

In this period, you will be given pure cultures of *E. coli*, *Proteus vulgaris* and *Enterobacter aerogenes,* all of which can be causative agents of kidney disease. You will inoculate these cultures into a variety of differential media, and next period, you will make observations of these reactions to learn the characteristics of these organisms. You will also be given an opportunity to perform serological typing of killed *Salmonella* cultures. Keep in mind that this serological typing is important not only for identification but also for epidemiological studies of the organism.

Materials

1. One to two ml of cultures of *E. coli, Proteus vulgaris* and *Enterobacter aerogenes*
2. Three plates of MacConkey's agar, three tubes of methyl red-Voges-Proskauer (MRVP) broth, four tubes of Kligler's iron agar, and four tubes of urea broth
3. Turbid suspensions (1 ml each) of killed cells of *Salmonella* strains for which antisera are available.[†] (At least two strains of *Salmonella* should be used.)
4. Antisera against the *Salmonella* strains; approximately three drops will be needed per student
5. Toothpicks for mixing cells and antisera

Procedure

1. Prepare smears of *E. coli, Proteus vulgaris* and *Enterobacter aerogenes* for staining in step 3. Allow to air dry.
2. Streak each of the three cultures onto a plate of MacConkey's agar. Inoculate each culture into one tube of MRVP and one tube of urea broth using a loop. Inoculate each culture into Kligler's iron agar using a needle to stab into the deep butt of the agar slant, then inoculate the slant by pulling the needle over the surface. Retain one tube of Kligler's iron agar and one tube of urea broth as controls. Incubate all plates and tubes at 37°. Observations of the urea broth and Kligler's iron agar should be done after 24 hours; observations of the MacConkey's agar and MRVP broth reactions can be made after 2 days.[‡]

   After 24 hours of incubation, observe the urea broth for urease activity. A positive reaction will be shown by a bright red color due to the formation of ammonia. (The pH indicator, phenol red, is red at alkaline pH and yellow at acid pH.) Organisms lacking the enzyme urease will give a yellow reaction in the broth.

   Also after 24 hours of incubation, observe the Kligler's iron agar for typical enteric reactions. These reactions are discussed more fully in Student Supplement 2. Reactions to be aware of are the color of the slant and butt, where yellow indicates acid production, orange indicates no reaction, and red indicates production of alkali. Compare the color of the culture to that of the uninoculated (orange) control to make clear the distinction between orange and red. Black indicates the formation of H$_2$S. Since enterics all ferment glucose, the butt of all three culture tubes should show an acidic reaction, although the production of H$_2$S may obscure this. If the organism can also ferment lactose, the slant will also be acidic; if it cannot ferment lactose, the acids produced from the glucose fermentation will be insufficient to overcome the alkaline re-action due to production of ammonia from the *oxidative* deamination of proteins on the slant, so the slant will show a net alkaline reaction. You will note that since this medium permits detection of a variety of bio-chemical reactions, it can be very useful in making quite rapid tentative identification of an organism.

3. Perform a Gram stain on the smears made earlier. Observe each smear for small, Gram-negative rods.

4. Add one drop of each culture of *Salmonella* to opposite sides of clean slides. To each drop, add a drop of antiserum against one of the strains. Mix each side with a clean toothpick, and observe for agglutination by gently tilting the slide. It is often easier to see agglutination against a dark background. Record the results. Now proceed to test the other antiserum against both strains of *Salmonella*. Note the specificity of re-sults.

Period B

In this period, you will finish observation of MacConkey's agar and perform the methyl red and Voges-Proskauer tests on the known cultures. You will also be given a mixture of two or three of the known cultures in which one of the organisms will be predominant. Your task will be to separate out the predominant culture and identify it. This will give you an idea of the clinical microbiologist's task in isolation and identification of disease-causing agents.

Materials

1. Plates and tubes from last period
2. Three empty tubes for the Voges-Proskauer test
3. Three pipettes
4. Methyl red, α-naphthol, and 40% KOH reagents
5. A mixture of two or three of the known organisms in which one of the organisms predominates by a ratio of 5–10:1**
6. One plate of MacConkey's agar

Procedure

1. Observe the plates of MacConkey's agar. MacConkey's agar selects for enterics by inhibiting organisms that cannot grow in the presence of bile

salts. The presence of crystal violet inhibits Gram-positive organisms such as *Streptococcus faecalis*, which would be able to grow in the presence of bile salts and are likely to be present in many samples of intestinal origin. Nonlactose-fermenting enterics will form colorless colonies. Lactose-fermenting enterics will form brick-red colonies surrounded by a halo of precipitate. Both of these effects result from the formation of acidic endproducts which cause the pH indicator neutral red to change to red and which precipitate the bile salts. Record the results.

2. Pipette one half of the MRVP broth from each tube into an empty, labelled tube for the Voges-Proskauer test. To each of the original tubes, mix in 2–3 drops of methyl red. A red color is a positive test, indicating the production of high levels of acid which have lowered the pH to below 4.6. A yellow color is a negative test. Record the results.

    To the other tube of each culture, add one drop of α-naphthol and mix. Then add 1 ml (10 drops from a dropper) of KOH and mix well. The development of a pink color within 15–20 minutes is a positive test for acetoin, a neutral endproduct produced by some enterics. Record the results.

3. Streak the plate of MacConkey's agar with your unknown mixture. Incubate the plate at 37° for 1–2 days.

Period C

In the following periods, you will be working exclusively with one isolate from the MacConkey's agar plate streaked last period from the mixture of unknown cultures.

Materials

1. The MacConkey's agar plate prepared last period
2. One nutrient agar slant to prepare a stock culture
3. One tube of MRVP broth, two tubes of Kligler's iron agar (one to use as an uninoculated control), and one tube of urea broth

Procedure

1. Observe the MacConkey's agar plate prepared last period for typical colonies of *E. coli*, *P. vulgaris* and *E. aerogenes*. Pick a *well-isolated* colony of the predominant type and use it as the inoculum source for the nutrient agar slant, MRVP broth, Kligler's iron agar, and urea broth. The slant for the stock culture should be inoculated first. Why? Incubate all media at 37°. Observations of the Kligler's iron agar and urea broth should be made after one day.

2. Prepare a Gram stain of the predominant colony type to be sure that your isolate is a Gram-negative rod.

In this period, you will be completing the identification of the predominant organism in the mixture, then performing an antibiotic sensitivity test to determine what the appropriate antibiotic therapy would be if this organism were a clinical isolate. The antibiotic sensitivity test can be used to screen the isolate for sensitivity to a large number of antibiotics with a minimum of manipulation and expense.

Materials

1. Plates and tubes inoculated last period
2. Methyl red, α-naphthol, and 40% KOH reagents
3. One empty tube and pipette for dividing the MRVP broth
4. One tube of top agar, melted and cooled to 45–50° in a waterbath
5. One plate of nutrient agar
6. Antibiotic sensitivity disks, 6–8 different disks per student; suggested antibiotics to use are given in Table 22-1, along with information for interpreting results

Table 22-1 Antibiotic susceptibility determinations for enteric bacteria

Antibiotic	Disk potency*	Zone of inhibition in mm for		
		Susceptibility	Intermediate resistance	Resistance
Ampicillin	10 μg	14 or more	12–13	11 or less
Bacitracin	10 units	13 or more	9–12	8 or less
Chloramphenicol	30 μg	18 or more	13–17	12 or less
Erythromycin	15 μg	18 or more	14–17	13 or less
Penicillin G	10 units	22 or more	12–21	11 or less
Polymyxin B	300 units	12 or more	9–11	8 or less
Streptomycin	10 μg	15 or more	12–14	11 or less
Sulfadiazine	300 μg	17 or more	13–16	12 or less
Tetracycline	30 μg	19 or more	15–18	14 or less

*These disk potencies are the standard quantities in most commercial disks for antibiotic susceptibility tests.

Procedure

1. Remove the top agar tube from the waterbath and inoculate it with cells from the nutrient agar slant. Mix the cells well but avoid bubbles, then pour the agar as an overlay on the nutrient agar plate. Be sure to tip the plate to ensure even distribution of the overlay. Set the plate aside to solidify.
2. Divide the MRVP broth and perform the methyl red and Voges-Proskauer tests. Record the results.
3. Based on the results of the reactions on MacConkey's agar, Kligler's iron agar, in urea broth, and the methyl red and Voges-Proskauer tests, identify the predominant organism in the unknown mixture

4. Add the antibiotic disks to the overlay plate, as illustrated in Figure 22-1. If the disks are to be added individually, use a sterile forceps. The disks should be evenly distributed over the plate so that results can be interpreted easily next period. Alternatively, a disk dispenser can be used which will add the disks quickly, aseptically and evenly spaced. This is the usual method of application for any clinical situation. Invert the plates and incubate at 37° for 2 days.

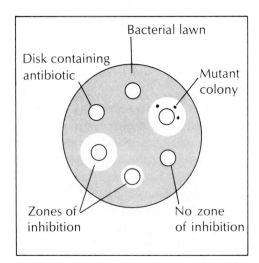

Figure 22-1 *Measurement of antibiotic susceptibility. Where there is no zone of inhibition the organism is resistant to that antibiotic. Colonies growing within a zone of inhibition have arisen from resistant mutants.*

Period E

Materials

1. The antibiotic sensitivity plate inoculated last period
2. A millimeter ruler

Procedure

Observe the plate for zones of inhibition around the disks containing antibiotics as illustrated in Figure 22-1. Measure the diameter of these zones and record the results. Rate the organism as susceptible, intermediate, or resistant to each antibiotic as indicated in Table 22-1. Which antibiotic would be appropriate for use in therapy? Be aware of the possible presence of colonies *within* the zone of inhibition which may be resistant mutants. If you had a choice of two antibiotics for therapy based on the size of the zone of inhibition, but one zone contained these resistant colonies, which antibiotic would be more appropriate (more likely to be successful) for therapy?

Questions

1. What is the basis of the agglutination reaction?
2. Why is serological identification of pathogenic isolates sometimes necessary?
3. Review the media used in this exercise. What are the selective and differential agents in MacConkey's agar? Kligler's iron agar? Urea broth?
4. Broadly outline the isolation and identification steps a clinical microbiologist might perform after being given a specimen from a diseased patient.

References

Lennette, E.H., A. Balows, W. J. Hausler and J.P. Truant, editors. 1980. **Manual of Clinical Microbiology**. 3rd edition. American Society for Microbiology. Washington, D. C. This manual contains a number of chapters on the isolation and identification of pathogenic enterics. It also contains a complete description of many of the tests and media used in this exercise.

*Serological typing uses specific antibodies that have been previously prepared by injecting an **antigen** into an animal. An antigen is a substance which causes the formation of antibodies. The outer surface of a bacterial cell (or any other cell) is covered with components which are antigenic. These components include the many proteins found in pili and flagella, the carbohydrates of the lipopolysaccharide of Gram-negative cells or of the teichoic acids of the Gram-positive cells, and any other external components such as capsules. The antibody specificities of these antigens can vary from one strain of bacterium to another.

†Both agglutinating *Salmonella* antigen preparations and antisera against these antigens are available from various commercial sources.

‡The Kligler's iron agar and urea broth can either be incubated for 24 hours, then observed, or they can be refrigerated after the incubation period until observations can be made.

**This mixture is best prepared by diluting overnight cultures of each organism by 1/10 in sterile saline, then adding appropriate amounts together to give the 5–10:1 ratio. This first 1/10 dilution prevents the metabolic endproducts of one organism from affecting another between the time of mixing and the time of use.

Notes

Name _____

Date _____

Report Sheet 22

Microbes of the Body: The Enterics

Serological typing of killed *Salmonella*

Salmonella strain	Antiserum type
	_____ _____

1. _____
2. _____

Give two reasons for the serological typing of pathogens in clinical procedures.

Characterization of enterics

Fill in the following chart

Test or characterization	*Escherichia coli*	*Proteus vulgaris*	*Enterobacter aerogenes*	Predominant organism in unknown
Colony morphology on MacConkey's agar				
Methyl red test				
Voges-Proskauer test				
Kligler's iron agar observations slant butt				
Urease				
Gram reaction				
Cellular morphology				

What is the predominant organism in the unknown? _____

(Continued on next page)

177

Antibiotic susceptibility testing

Fill in the following information about the antibiotic susceptibility of the predominant organism in your unknown.

Antibiotic	Zone of inhibition in mm	Level of susceptibility*	Presence of resistant colonies in the zone of inhibition
Ampicillin			
Bacitracin			
Chloramphenicol			
Erythromycin			
Penicillin G			
Polymyxin B			
Streptomycin			
Sulfadiazine			
Tetracycline			

*R = resistant; I = intermediate resistance; S = susceptible

Questions about the unknown

Answer the following questions about the predominant organism in your unknown:

Can it ferment glucose? How do you know?

Does it use a neutral endproduct or mixed-acid pathway in the fermentation of glucose?

(Continued on next page)

- Can it ferment lactose? How do you know?

- Is it a member of the family *Enterobacteriaceae*? How do you know? Are there any additional tests you might use to confirm this?

- What antibiotics could be used *successfully* in therapy if this organism were an isolate from an infected kidney?

Notes

Identification of Unknown Bacteria

23

In earlier exercises, you have studied a wide variety of bacteria, both Gram-positive and Gram-negative, rods and cocci, motile and nonmotile, fastidious and nonfastidious, aerobes and anaerobes. You have also performed a variety of tests for endproducts (for example, acetoin), enzymes (for example, amylase), and have observed various cellular structures such as capsules and flagella. You have also worked with a variety of selective and differential media that aid in the isolation and identification of bacteria. All of these characteristics of bacteria can be used as tools for the identification of unknown bacterial isolates. It is now time to assemble these tools for use. For this project, you will be given a culture containing a mixture of two or three organisms from the genera listed in Table 23-1. Each of the organisms in the mixture has been studied in a previous exercise. Your task will be to isolate pure cultures of all the organisms and identify each to genus. First, you will need to develop a procedure for the isolation of each organism and a plan for each identification that will involve the least amount of media and work. The easiest way to do this is to choose one general-purpose medium and two or three selective media which you will use for isolation. Additionally, you may wish to incubate plates at appropriate temperatures, for example 37° for enteric isolation plates, but 30° for plates used for organisms isolated from soil or water. When you observe your plates, be aware of differences in colony morphology, size, and pigment. Once you have obtained well-separated colonies on appropriate plates, you should prepare agar slants of each isolate. *Clear labelling is critical here.* These cultures can then be used to inoculate media for identification.

You should follow an identification scheme that will allow you to minimize the use of media. For example, if you perform tests for aerobiosis first and find your isolate is an obligate aerobe, then the use of glucose-Durham tubes or methyl red-Voges-Proskauer tests will not be necessary. (Why?) To prepare a scheme for identification, a list of critical tests is helpful; for example, list all you know about each genus in Table 23-1, then look for a

Table 23-1 Genera of organisms for use in the project of identification of unknown bacteria

Bacillus	*Neisseria*
Escherichia	*Proteus*
Klebsiella	*Pseudomonas*
Lactobacillus	*Staphylococcus*
Micrococcus	*Streptococcus*

few simple tests that will permit a rapid division of the genera into groups. Then, within each group, make a list of tests that can be used to identify each genus.

Period A

Procedure

Prepare a flow chart for the isolation and identification of the organisms listed in Table 23-1, using the media and reagents listed in Table 23-2. Note that the media and reagents available for this exercise will be confined to those listed in the table. You will have four *in class* laboratory periods in which to complete this exercise. Keep this in mind when developing your procedure. Your procedure should be checked by your instructor before beginning work. It is suggested that this check be done at least one laboratory period before beginning the work.

Table 23-2   Media and reagents available for the project on the identification of unknown bacteria

Media	Reagents
Plates	Gram staining reagents
All purpose Tween agar	5% H_2O_2
Eosin-methylene blue agar	40% KOH
MacConkey's agar	Kovac's reagent
Nutrient agar	Methyl red reagent
Sucrose-gelatin agar + NaN_3	α-naphthol reagent
Vogel-Johnson agar	Oxidase reagent
Tubes	Spore staining reagents
Glucose Durham tube	
Methyl red-Voges Proskauer broth	
Motility medium	
Tryptone broth	

Period B

Materials

1. An unknown broth mixture for each student. These mixtures should contain two or three organisms (the number to be determined by the instructor and announced to the student).*
2. Enough media should be available for each student to streak one plate of each selective medium and two plates of nonselective medium.

Procedure

Streak the unknown mixture onto one plate of each selective medium and two plates of the nonselective medium. Incubate each selective medium at a temperature appropriate for the organisms for which it is selective. (What would that be? Check back in the exercise in which you used the medium.) Incubate one of the plates containing a nonselective medium at 30° and the other at 37°. Begin Period C in either one or two days.†

Period C

Materials

1. Enough media should be available for each student to prepare agar slants of each isolate. If the mixture contains three organisms, each student should be supplied with five or six agar slants.
2. Five or six plates of a rich medium such as nutrient agar plus glucose to streak out the isolated colonies to check for purity and to perform catalase and oxidase tests.

Procedure

Observe the plates streaked in the last period. Choose colonies which appear to be different. Remember that the colonies present on the selective media will also be present on the nonselective medium, but will appear to be different. The purpose of the nonselective medium is to obtain cultures of any unknowns which cannot grow on the selective media. Prepare slants from isolated colonies. Also streak a plate from each of the same colonies used to prepare slants. Incubate both slants and plates at appropriate temperatures (the same as the temperature at which the plate used for isolation was incubated).

Period D

Materials

1. Enough media should be available to inoculate appropriate tests.‡
2. 5% H_2O_2 and oxidase reagent

Procedure

1. With colonies from the plates, perform the Gram stain, catalase test, and oxidase test on each of the isolates. If the plates appear to contain more than one colony type, proceed to re-isolate pure cultures as you did in Period C. Do not continue to step 2 until pure cultures are obtained!
2. Based on the Gram reaction and the catalase and oxidase tests from step 1, set up tests to identify the isolates to genus, choosing the media and reagents from those listed in Table 23-1. Incubate all cultures at appropriate temperatures.

Period E

Materials

Reagents should be available to perform any required tests.**

Procedure

Complete the identification of the isolates to genus.

*To prepare unknown mixtures, dilute overnight cultures of each organism 1/10 in sterile saline. Then mix these diluted cultures together in various combinations. In a mixture of Gram-negative and Gram-positive organisms, the Gram-positive organisms should outnumber the Gram-negative 10–15 to 1, or it will be extremely difficult to recover the Gram-positive organism. (This is probably due to the production of toxic endproducts by the Gram-negative organisms.) The mixtures should be prepared immediately before class to minimize cell death due to antagonistic interactions between cultures.

†Depending on teaching help and laboratory times available, it might be easiest to allow each student to work on this exercise at his or her own pace. For many students, it is possible to perform the steps of Period B one day, Period C the next, and Period D the following day. If possible, allow students access to the laboratory as needed. This gives them a better idea of the real timing involved in the isolation and identification of unknown organisms.

‡The approximate amount of each medium needed can be determined by looking at the number of organisms that might be tested in that medium and the number of times each of those organisms was used in an unknown mixture.

**It may also be necessary to make arrangements for any students who require more time to complete the identification.

Identification of Unknown Bacteria

For your unknown, fill in the following information:

Unknown code number: _____

Test or characteristic	Unknown isolate number					
	1	2	3	4	5	6
Gram reaction						
Cellular morphology						
Catalase						
Oxidase						
Other tests 1.						
2.						
3.						
4.						
5.						
6.						
7.						
8.						
Genus identification						

General Laboratory Procedures

Labelling of cultures

In laboratory work, it is very important to keep accurate records. Many cultures may be handled during any given period and they may often pertain to two or more different exercises. For this reason, the labelling of culture tubes, plates, slides and other material is very important. Use whatever system that seems convenient to you, but here are a few tips:

- Always label the cultures with markings that will not wash or rub off too easily. Red wax pencils are often used. They are inexpensive, don't require capping to prevent drying out between use, and come off easily during dishwashing. The disadvantages of using wax pencils are that they usually make a broad line, the writing can be rubbed off, and the marking is not heat-resistant. However, the alternatives to wax pencils also have disadvantages. Water-resistant felt-tip markers are easier to write with and the marks survive rubbing and hot water, but the markers are more expensive, require capping between use to prevent drying out, and must be removed from glassware with a solvent like ethanol before dishwashing. Tape or paper labels are very useful for long-term labelling but must also be removed prior to dishwashing.
- Always label the container *before* inoculation.
- Label the container *completely*. The label should indicate the culture name (e.g. for *Staphylococcus epidermidis* you might use "S.e."), the exercise number, the date of inoculation, and your name.
- If you *have* used water-resistant markers, tape, or paper labels on glassware, be sure to remove them before putting the glassware in the discard for sterilization and washing.

Record-keeping

To help keep accurate records, this manual has Report Sheets for each exercise to record pertinent observations. Questions have also been included in the procedure sections of the exercises. These questions rarely have a space for answering in the Report Sheet, as they are meant to be guides for proper observations. However, it will probably be useful to you to jot down your thoughts, ideas, or questions in the margin or in a separate notebook.

Sterilization of loops and other utensils

Safety Caution!

To sterilize loops and needles, hold the handle as you would a pencil and place it in the flame as indicated in Figure 1-2 in Exercise 1. It is important that the wire be heated red-hot. The part of the handle of the loop or needle that will enter the tube or plate should also be passed lightly through the flame to burn off dust. *The handle will not be sterile.* Allow the loop or needle to cool for 10 seconds before using it, both to avoid killing the cells and to avoid spattering which can lead to the production of contaminating aerosols. Forceps and glass spreaders cannot be sterilized by heating to red-heat. Dip the end to be sterilized into 95% ethanol, touch the wet end to a flame, and allow the alcohol to burn off *outside the flame.* Be careful that the burning alcohol does not run down to your hand or drip onto the laboratory bench. For complete safety, it is advisable to have the alcohol in a container with a cover, so that if the alcohol in the container accidentally catches on fire, the cover can be put on and the flame quickly extinguished. A forceps or spreader sterilized once can be used immediately without cooling, but if it is sterilized over and over, it will become too hot. Therefore, if you need to re-use the equipment, it is advisable to have more than one piece to allow alternation for cooling. For example, two spreaders can be available in the alcohol, with the flame-sterilization and use of them alternated.

Preparation of slides

The slide itself should be clean and grease-free so that wet mounts and smears which are made spread out evenly. For many uses, precleaned slides available commercially are clean enough, but sometimes even these need cleaning. The best method of cleaning is the use of a nonabrasive soap such as Bon-Ami. Wet the soap bar, rub some soap on the slide and allow it to dry. After it is dry, brush off the soap with a dry cloth (preferably not paper towel, as it leaves paper fibers on the slide). Microscope coverslips can be cleaned, if necessary, in a similar manner. You are cautioned, however, that coverslips break easily, and the shards of glass can easily become embedded in your skin. For that reason, coverslips are rarely re-used.

A faster cleaning method, effective only if the slide is fairly clean already, is to hold the slide in the flame for *one second* to burn off the dirt. Be sure to hold the slide with a forceps or clothespin when flaming. Also be sure that the slide is not heated for longer than one second, as the glass could shatter.

Safety Caution!

Wet mounts

Wet mounts are used for observing live microorganisms. It is important that all excess water be removed so that the coverslip does not float on the slide. To prepare a wet mount from a liquid culture, simply remove a small drop of the culture with a sterile loop or pipette (loop preferred), put the drop on the slide, and place a coverslip over the drop. To remove excess liquid, blot the top of the slide with blotting paper or paper towel or, alternatively, touch the paper to the side of the coverslip. If the culture is potentially pathogenic, this blotting paper should be decontaminated before discarding.

To prepare a wet mount from a colony on agar, place a small drop of water on the slide, then use a sterile needle to remove some of the cells and suspend them in the water. The suspension should appear barely turbid, not milky. Usually proper turbidity can be obtained if the needle is just touched to the colony, not scraped through it. With tiny colonies, a loop should be used instead of a needle. Add the coverslip and blot as described above.

To prepare a wet mount to view over a long period of time, the coverslip should be sealed to prevent evaporation of the liquid. This can be done with vaseline or nail polish. **Caution**. Either of these will strongly interfere with viewing if they get on the lens. Be careful! To seal a wet mount after it is prepared, cover the edges of the coverslip:slide interface with either vaseline (use a toothpick to apply) or nail polish. Be sure to let the nail polish dry *completely* before using the slide, to prevent possible contamination of the lens with nail polish. *A better alternative* is to seal the slide as it is prepared. To do this, smear a very thin layer of vaseline on the edge of the palm of your hand. Next, take the coverslip and scrape a small amount of the vaseline onto each edge of the coverslip. Place the coverslip on the slide *vaseline side down*. The advantage of this method is that the possibility of contaminating the lenses of the microscope with vaseline is very small. To remove vaseline or nail polish from the lens, clean the lens well with dry lens paper, then with lens paper moistened with distilled water, then with dry paper again. If this is not sufficient, *do not attempt any further cleaning without help from the instructor*! The instructor can *carefully* cleanse the lens with xylol to remove vaseline or with toluene to remove nail polish. Since both of these solvents dissolve the adhesive holding the last lens in the objective, it is important that the instructor perform this step.

Hanging drop slides

Hanging drop slides are often useful when the liquid medium contains large cells (such as protozoa) which would prevent proper contact between a slide and coverslip. To prepare a hanging drop slide, place a drop of the liquid culture on a *coverslip* placed on the lab bench. Next, add water, oil or vaseline in a thin line around the edges of the depression in the hanging drop slide.

Alternatively, the water, oil or vaseline can be added just at three or four spots around the depression, but remember, this may not completely seal the slide, so evaporation may be rather rapid. Next, invert the hanging drop slide onto the coverslip, centering the drop in the depression. Quickly and carefully, turn the slide right-side-up. The coverslip will adhere to the slide and the drop will now be hanging in the depression. (See Figure 2-2, Exercise 2.) Since the working distance between the objective and the slide is less than with a normal slide, special care must be taken to prevent damage to the objective lenses when higher magnifications are used.

Slides for staining

To prepare a slide for staining, follow the procedures for preparing a wet mount, *omitting* the addition of the coverslip. The proper density of cells on this smear is critical for proper staining. The density is correct if you can *just barely* observe turbidity in the drop. (If the drop is milky white, there are too many cells.) Spread the drop out on the slide to form a thin film. If the slide is clean, the liquid will spread evenly. Allow to dry.

Heat-fixing a slide for staining

Heat-fixing a slide helps cells adhere to the slide during the staining procedure. The exact mechanism at work here is unknown. Heat-fixing also kills the cells, which is desirable if the culture is potentially pathogenic. To heat-fix a slide, pass an air-dried slide (specimen side up) through a flame a few times. The slide should become warm but *not* hot. Too much heat will distort the cell structure. To practice correct heat-fixing, take a clean slide and pass it through the flame *quickly*. Touch the slide to your hand to feel how warm it is. Still cool? After a minute or so, pass it through the flame quickly two times. Feel the slide. Still cool? Increase the number of passes (or move the slide more slowly) until *you* know how many passes it takes to get the slide warm, but not hot. Be careful not to burn your hand. If the slide radiates enough heat that you feel it before it touches your hand, it is probably hot enough to burn your skin.

Safety Caution!

Staining procedures

Staining should always be done over a sink or beaker to catch any spills. It is also advisable to have the necessary staining reagents handy and in dropper bottles for easy application. The slide should have been heat-fixed (see above), unless otherwise stated, and should be handled with a clothespin to prevent staining your hands. Directions for preparing the staining reagents are given in Student Supplement 2. In all cases, the staining reagents should be used in just sufficient quantity to cover the smear to be stained. Do not flood the whole slide with staining reagent! When rinsing with water is indicated, the water should pass over the slide in a gentle stream. Blot carefully (no rubbing) so that cells are not removed from the slide.

The simple stain A simple stain is used when observations only on the morphology of the cells are needed. A variety of dyes can be used for a simple stain; the most commonly used dyes are crystal violet, methylene blue and safranin.

1. Flood the smear with the staining reagent. Allow the dye to act for 30 seconds to one minute.
2. Rinse the dye off in a gentle stream of water.
3. Blot the slide dry, then allow for complete drying in air before use.

The Gram-stain — rapid method The rapid method is faster and more reproducible than the conventional Gram-stain procedure given below. However, the longer method might be more manageable for the beginning student.

1. Stain the cells by flooding the smear with Gram's crystal violet.
2. Add 3–8 drops of 5% sodium bicarbonate ($NaHCO_3$) solution. Allow to act for 5–10 seconds.
3. Rinse the slide with Gram's iodine (do not rinse with water first), then flood the smear with the iodine. Allow the iodine to act for 5–10 seconds.
4. Decolorize by rinsing the slide with alcohol-acetone, then quickly flooding it with alcohol-acetone. Allow to act 5–10 seconds. Pour off the excess reagent and allow the slide to dry for 15 seconds. This decolorization step is the critical one, as it is in this step that Gram-negative cells lose the crystal violet dye. If this step is too short, the dye will be retained, giving Gram-negative cells the appearance of Gram-positive cells. If the step is too long, the dye will be lost from even the Gram-positive cells.
5. Counterstain by flooding the slide with safranin and allowing to act for 10 seconds. Rinse the slide with water and blot dry.
6. Observe the preparation with the 100× lens. Gram-positive cells will appear purple; Gram-negative cells will be pink or red. A more detailed discussion of the Gram-stain is given in Exercise 3 and in your textbook.

The Gram-stain — conventional method
1. Stain the cells by flooding the smear with Gram's crystal violet and allowing to act for 30 seconds. Rinse the slide with water.
2. Flood the smear with Gram's iodine and allow to act for one minute. Pour off the iodine and flood the smear again. After another minute, rinse off the iodine with water. Blot dry.
3. Decolorize by holding the slide at an angle over the sink or beaker and dripping alcohol-acetone over the smear, allowing the drips to catch on the edge of the slide briefly before falling off. Observe these drips. As soon as the drips lose the faint touch of blue, *rinse the slide with water immediately*. The decolorization step should not have taken more than 15–20 seconds.
4. Counterstain by flooding the smear with safranin and allowing to act for one minute. Rinse the slide with water and blot dry.
5. Observe the preparation with the 100× lens. Gram-positive cells will appear purple; Gram-negative cells will be pink or red.

The negative stain For a successful negative staining procedure, the proper India ink must be used. Some brands of India ink have very large carbon particles and are not appropriate for use. Generally, those inks prepared for rapid-flow graphic pens are the most successful. The negative stain is used primarily to reveal capsules. Remember that not all organisms which produce capsules do so under all cultural conditions.

1. Prepare a drop of culture as you would for any smear, but *do not* let it dry.
2. Add a small drop of India ink next to the drop of culture and allow the two drops to mix.
3. Place a coverslip over the liquid and blot as you would a wet mount.
4. Observe the preparation for capsules, looking in the area where ink and culture have mixed. If the area appears too dark, look on the other, lighter side of the slide. If the area appears too light, look on the darker side of the slide. Cells should be visible among the carbon particles. Cells with capsules will appear larger with the cell itself visible in the center of the capsule.

The spore stain The spore stain requires heating to impregnate the spores with dye. To accomplish this, a boiling water bath should be prepared over which a staining rack can be placed.

Safety Caution!

1. Place the slide on the staining rack above the boiling water bath. Flood the smear with malachite green solution. Heat the slide for 5 minutes, replacing any liquid that evaporates with fresh dye. Do not let the slide become dry or handle the hot slide with your fingers.
2. Allow the slide to cool to room temperature, then rinse with water.
3. Flood the smear with safranin, and allow it to act for 1–2 minutes. Rinse the slide with water and blot dry.
4. Observe the slide with the 100× lens. Spores will be green and vegetative cells will be pink or red.

The acid-fast stain This staining procedure requires heating to impregnate the cells with dye. To accomplish the heating, a boiling water bath should be prepared over which a staining rack can be placed.

Safety Caution!

1. Place the slide on the staining rack above the boiling water bath. Flood the smear with carbol fuchsin solution. Heat the slide for 5 minutes, replacing any liquid that evaporates with fresh dye. Do not let the slide become dry or handle the hot slide with your fingers.
2. Allow the slide to cool to room temperature, then rinse the excess carbol fuchsin off with water.
3. Flood the smear with acid-alcohol to decolorize those cells which are not acid-fast. Allow the acid-alcohol to act until the area of the slide containing the smear becomes light pink, then rinse the slide thoroughly with water.
4. Flood the preparation with methylene blue, and allow it to act for 2 minutes. Rinse with water and blot dry.
5. Observe the smear with the 100× lens. Acid-fast cells will be pink; non-acid-fast cells will be blue.

The flagella stain Cultures for use with the flagella stain must be handled carefully to prevent loss of the flagella. Also, the cells should be grown so as to maximize the production of flagella. Usually this means using a young culture.

1. Culture the organism on a slant of appropriate medium, such as nutrient agar or nutrient agar + glucose, incubating overnight at the optimum growth temperature.

2. Using a sterile *and cooled* loop, remove some of the cells from the slant. Touch the loop to the surface of a large drop of water in a watch-glass or in the depression of a hanging drop slide. Allow the organisms to float from the loop into the water. The suspension should be cloudy. *Do not agitate the water with the loop.* Allow this drop to sit for 5 minutes. (Note: this length of time is sufficient to dry a small drop of water, so be sure the drop of water is sufficiently large to avoid this.) Now carefully touch the top of this drop of water + cells with a cool loop and transfer a loopful of cells to a clean, glass microscope slide. Slides from commercial slide boxes are *not* clean enough for this technique, as the oils from manufacturing will cause the flagella to align closely with the edge of the cell as the smear dries. Slides should be cleaned with soap such as Bon-Ami, then, a few minutes before use, passed through a flame with the specimen-side toward the flame.

    Allow this smear to air-dry. Heat-fixing is not necessary as the flagella-staining reagents will chemically fix the cells to the slide.

3. Flood the smear with fresh flagella stain (see Student Supplement 2 for preparation), and allow it to act for 1–2 minutes. Rinse with a gentle stream of water and allow to air-dry.

4. Observe the slide for flagella using the 100× lens. These will be seen as small hair-like fibers extending from the cell. The motile enteric bacteria, such as *Escherichia* or *Proteus*, will have peritrichous flagella, that is, flagella distributed all around the cell surface. *Pseudomonas* will have polar flagella. Even well-stained flagella are difficult to see. Carefully cleaned microscope lenses will help.

Enumeration of microorganisms by plate count

Either pour plates or spread plates can be used to count the number of viable organisms in a sample. In the pour plate, the culture is mixed with the agar before it has solidified, and the colonies which form are embedded in the agar. In the spread plate, the culture is spread over the surface of the agar plate after the agar has solidified and dried. Spread plates are more useful if the cells to be enumerated are heat-sensitive. Pour plates are easier for the beginning student to prepare, although counting of the colonies is more difficult.

Pour plates

Materials include sterile empty petri plates and molten agar in either tubes or a bottle. The agar should be cooled to 45–50° in a water bath. If tubes are used, each tube should contain 15–25 ml of agar.

1. Label the sterile petri dishes on the bottom.
2. If tubes of agar are used, inoculate the agar directly with the appropriate aliquot of the diluted culture and mix well by swirling. Wipe the tube

to remove excess water, flame the lip of the tube, and then pour the agar into the plate, tilting the plate slightly or swirling it on the laboratory bench to ensure that the bottom of the plate is uniformly covered with agar.

3. If bottles of agar are used, place the inoculum in the empty plate. Wipe the water off the bottle, flame the lip of the bottle, and then pour agar into the plate until the plate is approximately one-half full. Carefully swirl the plate to mix the inoculum and agar; avoid splashing the agar onto the lid of the plate.

4. Allow the plates to solidify *undisturbed* . Invert the plates for incubation.

Spread plates

Materials for the spread plate procedure include plates of medium that are solidified and have been dried for one or two days at room temperature, a glass rod that has been bent in the shape of a hockey stick to spread the inoculum, and 95% alcohol to flame-sterilize the glass spreader.

1. Using a sterile pipette, transfer 0.1 ml of inoculum to the center of an agar plate.

2. *Before* the inoculum soaks into the agar, flame-sterilize the glass spreader, allow it to cool a few seconds in the air, then spread the inoculum in the plate to completely cover the agar surface. If the glass spreader spits as it touches the agar, it was too hot. Refer to the proper procedure for flame-sterilization of glass rods elsewhere in this supplement.

3. Invert the plates for incubation.

Streaking agar plates for isolated colonies

When agar plates are streaked to obtain isolated colonies, it is important that the agar surface be well dried. This can be accomplished by natural drying at room temperature for 2–5 days, or it can be accomplished quickly by placing the plate in a sterile hood or room with the bottom of the plate angled on the inverted top (see Figure SS1-1) for 1–2 hours. This second method is not useful for large numbers of plates and does have an increased chance of contamination.

Plates can be streaked using either the loop or a sterile swab. The loop is less expensive, but the swab often gives better results, especially if the agar is rather soft (because of a low percentage of agar in the medium, for example). Directions are given for both loop and swab for a three-phase streaking pattern. A four-phase pattern can also be used.

Streaking with a loop

1. Sterilize the loop and allow it to cool. Use a culture grown in either a liquid or solid medium. If a liquid medium is the source of cells, the loop should be inserted into the medium; then, when being removed

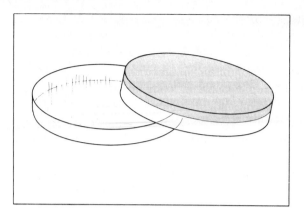

Figure SS1-1 *Procedure for rapid drying of plates for streaking. Note that* both *lid and bottom are* inverted.

from the container, the loop should be "broken" so that only those cells adhering to the wet wire and not those in a film within the wire are transferred. If the cells are obtained from a solid medium, the loop should be touched to the colony *very lightly*. You should *not* see a clump of cells on the loop. (Alternatively, a needle can be used to make the initial transfer of cells to the plate, switching to a loop for any streaking.)

2. Place the cells at one edge of the plate as illustrated in Figure 1-4, then streak the culture on approximately one-third of the plate, leaving no open spaces. The handling of the plate can be accomplished in a number of ways, all of which attempt to minimize possible contamination by either keeping the lid over the plate or by keeping the plate upside down (see Figure 1-5).

3. Sterilize the loop and allow it to cool in the air for 15–30 seconds. Touch the loop to an unused edge of the agar surface to cool it completely before continuing. Then pull the loop through one edge of the previous streaks *one or two times* to re-inoculate the loop. Remember, you are attempting to get isolated colonies at the end of all this, so you do not want the loop to have too many cells on it. Now streak the second third of the plate *avoiding the first third*. See Figure 1-4.

4. Repeat step 3, being sure that you obtain inoculum from the area most recently streaked.

5. Invert the plate for incubation. After incubation, there should be isolated colonies in one or more sections of the plate. If the plate has confluent growth, one of the following may have happened:

   • The agar surface was too wet, allowing the colonies to run together.
   • The plates were not incubated upside-down and water condensing on the lid dropped on the colonies.
   • The loop moved through too many of the previous streaks to obtain inoculum after the secondary sterilization of the loop.

If the plate is almost devoid of growth, either the original inoculum contained few cells or your loop was too hot. If one or two portions of the plate have considerable growth and the next portion is devoid of growth, the loop was too hot at that point.

Streaking with a cotton swab

1. Remove the sterile cotton swab from its wrapping, being sure not to touch the swab to anything. If a liquid culture is used, insert the swab into the liquid, then roll the swab on the side of the container to remove excess liquid. Using the swab like a loop, streak one-third of the agar surface as outlined for the loop above. Discard the swab, obtain a second (and then a third) swab and proceed to streak the plate as would be done with a loop. If a culture is obtained from a solid medium, a single swab can be used to do a complete streak plate from a colony provided the swab is handled carefully. To do this, lightly touch one edge of the swab to the colony. Then streak the first third of the plate with this material. Now rotate the swab to an unused portion and streak the second phase. Rotate the swab again to complete the procedure.

2. Invert and incubate the plates. After incubation, there should be isolated colonies in some area of the plate. If the plate is overgrown, either the plates were not properly dried or you forgot to either change swabs or rotate the dry swab to a sterile side. If the plate is almost devoid of colonies, the only possibility (assuming the cultural conditions were correct for growth on the plate) is that the original inoculum had very few cells in it. Notice how the use of swabs eliminates problems with hot loops, such as the killing of cells and the creation of aerosols. Swabs are also preferred when potential human pathogens are being streaked.

Decontamination procedures

Work in a microbiology laboratory often involves use of pathogens or potential pathogens for either animals or plants. For this reason, it is important to observe rather strict procedures for decontamination of the laboratory, equipment, and personnel. In addition, it is difficult to perform aseptic procedures in a laboratory that is not properly cleaned during and after use. The following list contains procedures which will minimize contamination during a laboratory period, as well as procedures for decontamination of the laboratory after use.

1. Wash hands well with soap and water both before and after laboratory work. Liquid soap that is in a wall-hung container is preferred to soap bars to avoid contamination of the soap by any highly contaminated hands.

2. Wipe the bench with a disinfectant such as Amphyl or Roccal before beginning and after completing work. Squirt bottles filled with disinfectant available at each bench will make the job much easier. Directions for preparation of working solutions of these disinfectants are given on the label.

3. All slides should be discarded into disinfectant or into a container which will be sterilized by autoclaving.

4. All glass tubes, bottles, flasks, pipettes, and plates containing cultures should be sterilized by autoclaving before dishwashing.

5. All plasticware, such as petri plates and pipettes, should be sterilized by autoclaving before discarding in the normal trash. Contaminated plasticware can be placed in a small metal trash bucket lined with an autoclavable plastic bag, and the whole bucket placed in the autoclave. After sterilization, allow the bucket to cool and the agar to solidify in the bag before discarding to the trash.

6. All loops, needles, and spreaders should be sterilized properly before placing them on the bench or in the storage area. Proper sterilization includes avoiding aerosol production from a loop by lowering the loop slowly into the flame, to allow any liquid to dry before heating the wire to red-heat.

7. Spills of cultures should be wiped up using first a dry paper towel, which is then discarded into the plasticware waste. The spillage area should then be wiped thoroughly using the bench disinfectant. The person cleaning the area should wash his hands well before resuming work.

8. If the atmosphere in the laboratory appears to be highly contaminated with microorganisms, as indicated by high contamination rates on plates, or if aerosols are a possibility because of the procedure being used, contamination can be reduced by using a "wet" lab bench to trap the microorganisms. A "wet" bench is made by placing a towel saturated with disinfectant on the working area of the bench. Microorganisms will be caught and killed in the disinfectant. In addition, dust, which might contain microorganisms, is also caught. One caution: do not rest arms or hands on the moist towel as most disinfectants will irritate the skin after continuous exposure of this type.

Laboratory Tests

Amylase test

Amylase is an extracellular enzyme which hydrolyzes the polysaccharide starch into its constituent monosaccharides. The production of this enzyme by a microorganism can be detected on a medium containing starch. The culture should be streaked across the center of the starch-agar plate and incubated for 2–4 days. After incubation, the plate is flooded with iodine solution to detect the presence or absence of starch. Starch reacts with iodine to give a blue color. If the starch has been degraded near the colony, after the addition of iodine there will be a clear (non-blue) area around the colony. The presence of the clear area is a positive test for amylase production (but a negative test for the presence of starch). As the color production by the starch-iodine complex is transient, the plates should be observed immediately after the addition of the iodine.

Catalase Test

Catalase is an intracellular enzyme which catalyzes the breakdown of hydrogen peroxide (H_2O_2) to water (H_2O) and oxygen (O_2). The presence of the enzyme can be detected by adding a drop of 5% H_2O_2 to a colony and observing for the production of bubbles of O_2. This test can also be performed by putting a drop of 5% H_2O_2 on a slide and adding the cells to the drop. This latter method is required if the culture to be tested is growing on blood agar, since blood cells also contain catalase. The catalase test is useful for a quick separation of catalase-positive staphylococci from catalase-negative streptococci.

Coagulase Test

Coagulase is an extracellular enzyme which causes the formation of fibrin clots in plasma. It can be detected by adding 0.5 ml of a 1- or 2-day-old culture to 0.5 ml of rabbit plasma and incubating this mixture at 37°. An organism producing coagulase will cause the clotting (coagulation) of the plasma, usually in less than 2 hours. This test is usually done as a diagnostic step in the identification of pathogenic *Staphylococcus aureus*.

β-Galactosidase Test

β-Galactosidase is an intracellular enzyme which hydrolyzes lactose to glucose and galactose. The substrate (lactose) and the products (glucose and galactose) of this enzyme in the cell are difficult to assay, so that a modified substrate, ortho-nitrophenyl-β,D-galactoside (ONPG), is used instead. β-Galactosidase will cleave ONPG to galactose and ortho-nitrophenol, a bright yellow compound. To test for the enzyme, the cells in the culture must first be treated with toluene to make the cell membrane permeable to the substrate ONPG. This is done by adding 0.5 ml of 5% toluene in acetone to a 5 ml culture, shaking the culture vigorously, and allowing it to sit for 10 minutes before adding the ONPG. After the 10 minutes has elapsed, 0.5 ml of 1.3×10^{-3} M ONPG is added. If β-galactosidase is present, the yellow color of the ONP product will be apparent within 5–10 minutes.

Indole Test

See Tryptophanase Test.

Methyl Red Test

The methyl red test is used to detect the formation of large amounts of acids in the fermentation of glucose. It is done by adding 5 drops of methyl red reagent without mixing to 2.5 ml of a culture grown in methyl red-Voges-Proskauer broth. If the culture has produced large amounts of acids (usually because it uses the mixed acid pathway of fermentation), the pH will be low enough to change the color of the methyl red indicator from red to yellow.

Oxidase Test

The oxidase test determines the ability to oxidize aromatic amines to colored products. This can be used as a differential test for many different organisms. It is performed by adding a drop of oxidase reagent (1% tetramethyl-para-phenylenediamine) to a colony. If the colony quickly turns a deep purple, it is oxidase positive.

Serological Tests for Agglutination

Agglutination is a clumping of microbial cells due to the antibody:antigen reaction which occurs after antiserum has been added to a cell suspension on a slide. Place a drop of cell suspension on a slide, add a small drop of antiserum, and mix with a clean toothpick. Agglutination is often easily observed against a black background, but visualization may be improved by tilting the slide back and forth. A negative control should be performed using serum which has no antibodies against the cells being tested. This negative control allows distinction between a true agglutination reaction and a small amount of cell clumping that may occur as the drop loses moisture.

Starch Test

See Amylase Test.

Tryptophanase Test

Tryptophanase is an intracellular enzyme which catalyzes the breakdown of the amino acid tryptophan to pyruvate and indole. The pyruvate is used by the cell as a carbon and energy source; the indole is excreted as a waste product. To test for tryptophanase, a culture is grown in a medium containing high levels of tryptophan, such as tryptone broth. After 2–4 days of incubation, several drops of Kovac's reagent are added to the broth, but *not mixed in*. The development of a red ring at the top of the broth is a positive test.

Voges-Proskauer Test

This test is used to detect the formation of the neutral fermentation product acetoin (also known as acetylmethyl carbinol), produced from glucose by some organisms as an intermediate on the way to 2,3-butanediol. To perform the test, 0.6 ml of α-naphthol reagent and 0.2 ml of 40% KOH are added to 2.5 ml of a culture grown in methyl red-Voges-Proskauer broth. This mixture is shaken well and observed for the production of a pink-red color within 15–20 minutes.

Notes

Media and Reagents

This supplement contains the recipes for all the media and reagents used in the laboratory exercises. Following each recipe is a description of the preparation and use of the medium or reagent. Many of the media contain complex ingredients, such as peptones, extracts, and infusions. Peptones are obtained from the digestion of proteinaceous animal or plant materials and are rich sources of nitrogenous compounds, such as peptides, amino acids and ammonia. They also supply some vitamins and many required minerals (elements), such as sulfur, magnesium and phosphorus. Infusions are aqueous extracts of ground, defatted meats which supply nitrogenous compounds, minerals, and vitamins. Extracts are similar to infusions but may be from yeast, beef heart, or plant material. Extracts also supply nitrogenous compounds, minerals, and vitamins. Yeast extract is often used to supply B-complex vitamins. Since the chemical contents of peptones, infusions and extracts are not completely known, media containing these materials are labelled *undefined* or *complex*.

Agar, an impure polysaccharide extracted from certain marine algae, is used as the solidifying agent in most media because it is not digested by most microorganisms and has a useful temperature range of solidification. To liquify (or dissolve), agar must be raised to a temperature near boiling.* Once liquid, agar does not solidify again until the temperature drops to about 43°. For inoculating or working with liquid agar, it should be kept in a water bath at 45–50°, which is cool enough to allow inoculation of most microorganisms without cell death (see procedures for pour plates and agar overlays in the manual).

*This allows cultivation of thermophilic organisms on a solid surface and is one of the advantages of using agar instead of other solidification agents such as gelatin.

Media

All Purpose Tween (APT) Agar

Tryptone	12.5 g
Glucose	10.0
Yeast extract	7.5
Na_3 Citrate	5.0
NaCl	5.0
K_2HPO_4	5.0
$MgSO_4$	0.8
Tween 80	0.2
$MnCl_2$	0.14
$FeSO_4$	0.04
Thiamine HCl	0.0001
Agar	15.0
Distilled H_2O	1000 ml

Preparation   Available commercially in dehydrated form. Do not overheat.

Use   This medium is a rich medium used to isolate and cultivate fastidious organisms, especially the lactic acid bacteria. (The Tween 80 supplies a nontoxic source of oleic acid for these organisms.)

All Purpose Tween Agar + NaN_3

Ingredients as in All Purpose Tween Agar, to which 0.2 g of NaN_3 is added before autoclaving.

Preparation   All Purpose Tween agar is available

commercially in dehydrated form, to which NaN_3 can be added. Do not overheat.

Use   This medium is used for the selective isolation of the lactic acid bacteria. NaN_3 is a cytochrome inhibitor which prevents the growth of organisms with a respiratory system.

All Purpose Tween (APT) Broth

Tryptone	12.5 g
Glucose	10.0
Yeast extract	7.5
Na_3 Citrate	5.0
NaCl	5.0
K_2HPO_4	5.0
$MgSO_4$	0.8
Tween 80	0.2
$MnCl_2$	0.14
$FeSO_4$	0.04
Thiamine HCl	0.0001
Distilled H_2O	1000 ml

Preparation   Available commercially in dehydrated form. Do not overheat.

Use   This medium is a rich medium used to isolate and cultivate fastidious organisms, especially the lactic acid bacteria. (The Tween 80 supplies a nontoxic source of oleic acid for these organisms.)

Blood Agar

Beef heart infusion	500 g
Tryptose	10.0
NaCl	5.0
Defibrinated blood	50–100 ml
Agar	15.0 g
Distilled H_2O	1000 ml
pH 7.2	

Preparation   Blood agar base (containing all the ingredients listed above except the blood) is available commercially in dehydrated form. Sheep blood should be obtained *aseptically*. If obtaining sheep blood is difficult, rabbit or human (from blood banks) blood can be used. Rabbit or human blood will not, however, give typical hemolysis reactions. Alternatively, blood agar plates (prepared) are available commercially.

Use   This medium is used for the isolation and cultivation of nutritionally versatile human microflora. It is also used to detect production of hemolysins which lyse red blood cells either completely (β-hemolysis) or partially (α-hemolysis). β-hemolysis is indicated by a clearing of the medium; α-hemolysis by a greening of the red blood cells.

Bottom Agar

Peptone	0.6 g
Yeast extract	3.0
NaCl	2.0
Agar	15.0
Distilled H_2O	1000 ml

Use   This medium is used as the bottom layer of agar when a bacterial lawn will be grown in an agar overlay (for example, bacteriophage titering, antibiotic susceptibility testing).

Brain Heart Infusion Broth or Agar

Calf brains, infusion from	200.0 g
Beef heart, infusion from	250.0
Proteose peptone	10.0
Glucose	2.0
NaCl	5.0
Na_2HPO_4	2.5
Distilled H_2O	1000 ml

Preparation   Available commercially in dehydrated form. If a solid medium is needed, add 15 g of agar per liter.

Use   For the cultivation of more fastidious organisms.

Eosin Methylene Blue (EMB) Agar

Peptone	10.0 g
Lactose	10.0
K_2HPO_4	2.0
Eosin Y	0.4
Methylene blue	0.065
Agar	15.0
Distilled H_2O	1000 ml

Preparation   Available commercially in dehydrated form. Do not overheat.

Use   This medium is a selective-differential medium used especially for the differentiation of *E. coli,* coliforms, and non-lactose-fermenting, Gram-negative organisms. The eosin Y and methylene blue dyes inhibit the growth of Gram-positive organisms. Organisms such as *Klebsiella* and *Enterobacter,* which ferment lactose via the butanediol pathway (yielding neutral endproducts) will produce dark-colored colonies. This color may be only in the center of the colony or throughout. Organisms such as *E. coli,* which ferment lactose via the mixed-acid pathway, will produce sufficient acid to cause complexing of the two dyes, giving a green sheen to the colonies. This sheen is obvious with *reflected* light. Non-lactose fermenters will produce colorless or light pink colonies.

Gelatin Medium (Nutrient Gelatin)

Peptone	5.0 g
Beef extract	3.0
Gelatin	120.0
Distilled H_2O	1000 ml

Preparation   Available commercially in dehydrated form. This medium must be heated at 50° to dissolve the gelatin before dispensing into tubes. Usually the tube is filled with only 1/4–1/3 of its total volume.

Use   This medium is used to test for the production of

the extracellular enzyme gelatinase. This enzyme hydrolyzes gelatin into its constituent amino acids. A positive test for gelatinase is indicated by failure of the medium to form a solid gel after being cooled on ice for 10–15 minutes. Unhydrolyzed gelatin will gel in that period of time (see Gelatinase Test, Student Supplement 1).

Glucose Agar Deep

Tryptone	10.0 g
Yeast extract	3.0
Bromcresol purple	0.02
Glucose	10.0
Agar	12.0
Distilled H_2O	1000 ml
pH 7.4 ± 0.1	

Preparation   This medium can be prepared as listed above, with the dye weighed out. Alternatively, a stock solution of 2.0% bromcresol purple in 95% ethanol can be prepared for this and other media containing the dye and 1 ml/l added. Tubes must be melted and cooled to 45–50° for inoculation, then cooled to solidify for incubation.

Use   This medium is used to detect acid and gas production in the fermentation of glucose. Acid production will lower the pH of the medium, resulting in a change of the pH indicator, bromcresol purple, from purple to yellow. Gas will be evident by the breakup of the agar in the tube.

Glucose Durham Tube

Tryptone	10.0 g
Yeast extract	3.0
Bromcresol purple	0.02
Glucose	10.0
Distilled H_2O	1000 ml
pH 7.4 ± 0.1	

Preparation   As mentioned in Glucose Agar Deeps, the dye can be added from a stock solution rather than as a solid. The medium should be dispensed into tubes which contain a smaller inverted tube. Fill the larger tube about 3/4 full of medium. Autoclave. As the tubes cool after autoclaving, the small inverted tubes will fill with the broth.

Use   This medium is used to detect the production of acid and gas in the fermentation of glucose. Acid production will lower the pH of the medium, resulting in a change of the pH indicator, bromcresol purple, from purple to yellow. Gas will be evident as a bubble in the inverted tube. Tubes with negative results for gas production should be tapped lightly on the bottom and observed for very small gas bubbles rising to the surface. Those tubes showing these small bubbles should be recorded as weakly positive for gas production.

Heart Infusion Broth

Beef heart, infusion	500.0 g
Tryptose	10.0

NaCl	5.0
Distilled H_2O	1000 ml
pH 7.4 ± 0.2	

Preparation   Available commercially in dehydrated form.

Use   This medium is used for the cultivation of a broad range of bacteria.

Kligler's Iron Agar

Beef extract	3.0 g
Yeast extract	3.0
Peptone	15.0
Proteose peptone	5.0
Lactose	10.0
Glucose	1.0
$FeSO_4$	0.2
NaCl	5.0
$Na_2S_2O_3$	0.3
Agar	12.0
Phenol red	0.024
Distilled H_2O	1000 ml
pH 7.4	

Preparation   Available commercially in dehydrated form. Steam to dissolve, then dispense 10 ml into tubes. After autoclaving, allow the agar to solidify while the tubes are in a slanted position to form slants with deep butts. The medium should be orange, not yellow or red.

Use   This medium is used in the differentiation of enteric organisms. The tube is inoculated by streaking the organism on the slant, then stabbing the butt. After incubation, a variety of results can be expected. A black butt indicates the production of H_2S from either the thiosulfate ($Na_2S_2O_3$) or the sulfur-containing amino acids in the medium. This H_2S combines with the iron ($FeSO_4$) to form FeS, a black precipitate. This reaction often obscures other reactions in the butt of the tube. A yellow butt indicates the production of acid during the fermentation of lactose or glucose, whereas an orange or red butt indicates lack of fermentation. Enterics are expected to produce a yellow butt. Gas production can also be detected by the breakup of the agar in the butt.

Reactions that can occur on the slant are the production of acid from the fermentation of glucose or lactose and the production of base ($NH_4{}^+$) from the *oxidative* deamination of proteins. All enterics are capable of oxidative deamination of proteins. The acid production from the fermentation of the low levels of glucose in the medium is insufficient to counterbalance the base production from the deamination of proteins. Therefore, organisms that *do not* ferment lactose show a net basic reaction (red slant). Those organisms which also ferment lactose, a sugar present in much higher levels in this medium, produce enough acid to show a net acidic reaction (yellow slant).

This medium, then, can be used to indicate H_2S production, glucose fermentation, lactose fermentation, and oxidative deamination of proteins. Each of these characteristics can be used diagnostically in the identification of organisms.

Lactose Agar Deeps

Tryptone	10.0 g
Yeast extract	3.0
Bromcresol purple	0.02
Lactose	10.0
Agar	12.0
Distilled H_2O	1000 ml
pH 7.4 ± 0.1	

Preparation   As mentioned for Glucose Agar Deeps, the bromcresol purple dye can be added as a solid or as a solution. Tubes must be melted and cooled to 45–50° for inoculation, then cooled to solidify for incubation.

Use   This medium is used to detect acid and gas production in the fermentation of lactose. Acid production will lower the pH of the medium, resulting in a change of the pH indicator, bromcresol purple, from purple to yellow. Gas will be evident by the breakup of the agar in the tube.

Lactose Durham Tube

Tryptone	10.0 g
Yeast extract	3.0
Bromcresol purple	0.02
Lactose	10.0
Distilled H_2O	1000 ml
pH 7.4 ± 0.1	

Preparation   As mentioned in Glucose Agar Deeps, the dye can be added from a stock solution rather than as a solid. The medium should be dispensed into tubes which contain a smaller inverted tube. Fill the larger tube about 3/4 full of medium. Autoclave. As the tubes cool after autoclaving, the small inverted tube will fill with the broth.

Use   This medium is used to detect the production of acid and gas in the fermentation of lactose. Acid production will lower the pH of the medium, resulting in a change of the pH indicator, bromcresol purple, from purple to yellow. Gas will be evident as a bubble in the inverted tube. Tubes with negative results for gas production should be tapped lightly on the bottom and observed for very small gas bubbles rising to the surface. Those tubes showing these small bubbles should be recorded as weakly positive for gas production.

Lactose Lauryl Tryptose Broth

Tryptose	20.0 g
Lactose	5.0
K_2HPO_4	2.75
KH_2PO_4	2.75
NaCl	5.0
Na lauryl sulfate	0.1
(Na dodecyl sulfate)	
Distilled H_2O	1000 ml

Preparation   Available commercially in dehydrated form. Mix, then dispense into Durham tubes. (See notes on the preparation of Glucose Durham Tubes.) If large amounts of water are to be tested, double-strength medium should be used to avoid excessive dilution. To obtain double-strength medium, double all ingredients except the water. Double-strength medium should be distributed in 5 ml amounts; single-strength medium in 9 ml amounts.

Use   This medium is used to detect coliforms and fecal coliforms in water. The coliforms will produce gas (detected as a bubble in the small inverted vial) from the fermentation of lactose at 35° after 48 hours. Fecal coliforms produce the gas also at 44.5°. Nonenteric organisms will be inhibited by the presence of the sodium lauryl sulfate.

MacConkey's Agar

Peptone	17.0 g
Proteose peptone	3.0
Lactose	10.0
Bile salts	1.5
NaCl	5.0
Agar	13.5
Neutral red	0.03
Crystal violet	0.001
Distilled H_2O	1000 ml
pH 7.1 ± 0.1	

Preparation   Available commercially in dehydrated form either as MacConkey's Agar Base (as above without lactose) or as MacConkey's Agar (has lactose).

Use   This medium is used for the isolation and differentiation of the enterics. It is a selective-differential medium. The bile salts select against all nonenteric organisms, while the crystal violet selects against Gram-positive organisms, such as *Streptococcus faecalis*, which can grow in the presence of bile salts. Nonlactose fermenters will form colorless or light pink colonies; lactose fermenters will form brick red colonies surrounded by precipitated bile salts.

Methyl Red-Voges-Proskauer (MRVP) Broth

Buffered peptone	7.0 g
Glucose	5.0
K_2HPO_4	5.0
Distilled H_2O	1000 ml
pH 6.9 ± 0.1	

Preparation   Available commercially in dehydrated form.

Use   This medium is used to test for the production of high amounts of acid or specific neutral endproducts in the fermentation of glucose. High acid can be detected by a red reaction following the addition of methyl red, a pH indicator (see Methyl Red Test–Student Supplement 1). Specific neutral endproducts (acetoin, butanediol) can be detected as a pink coloration following addition of α-napthol and 40% KOH in the Voges-Proskauer Test (see Student Supplement 1).

Minimal Medium 1 (MM1)

K$_2$HPO$_4$	7.0 g
KH$_2$PO$_4$	2.0
(NH$_4$)$_2$SO$_4$	1.0
MgSO$_4$	0.1
Glucose	5.0
Agar	15.0
Distilled H$_2$O	1000 ml
pH 7.0 ± 0.2	

Preparation This medium should be prepared as two solutions for best results. Solution A contains the first three ingredients in 500 ml of distilled H$_2$O and is adjusted to pH 7.0 ± 0.2. Solution B would contain the last 3 ingredients in 500 ml of distilled H$_2$O. After autoclaving, the two solutions are mixed together aseptically and thoroughly before being poured into plates. The medium should not be brown. If prepared as one solution, the sugars (glucose and those in the agar) may caramelize in the presence of the phosphates and insoluble Mg phosphates may also be formed.

Use This medium is used as a minimal medium for the growth of *E. coli*. This medium supplies the major elements P, S, N and Mg with glucose serving as the organism's carbon and energy source. This medium need not be supplemented for the growth of wild type *E. coli*, but would require at least addition of iron (FeSO$_4$; 0.02 g/l) for most other organisms.

Minimal Medium 2 (MM2)

K$_2$HPO$_4$	7.0 g
KH$_2$PO$_4$	2.0
(NH$_4$)$_2$SO$_4$	1.0
Glucose	5.0
Agar	15.0
Distilled H$_2$O	1000 ml
pH 7.0 ± 0.2	

Preparation This medium should be prepared as two solutions for best results. The preparation is outlined with Minimal Medium 1.

Use This medium is used to demonstrate elemental requirements of organisms. It will *not* support the growth of *E. coli*.

Minimal Medium 3 (MM3)

K$_2$HPO$_4$	7.0 g
KH$_2$PO$_4$	2.0
NH$_4$Cl	1.0
MgCl$_2$	0.1
Glucose	5.0
Agar	15.0
Distilled H$_2$O	1000 ml
pH 7.0 ± 0.2	

Preparation This medium should be prepared as two solutions for best results. This preparation is outlined with Minimial Medium 1.

Use This medium is used to demonstrate elemental requirements of organisms. It will *not* support the growth of *E. coli*.

Minimal Medium 4 (MM4)

K$_2$HPO$_4$	7.0 g
KH$_2$PO$_4$	2.0
(NH$_4$)$_2$SO$_4$	1.0
MgSO$_4$	0.1
Glucose	5.0
Agar	15.0
Distilled H$_2$O	1000 ml
pH 7.0 ± 0.2	
Amino acid	40 mg

Preparation This medium should be prepared in three solutions. Solution A should contain the first three ingredients in 500 ml of distilled H$_2$O and have its pH adjusted. Solution B should contain the MgSO$_4$, glucose and agar in 500 ml of distilled H$_2$O. After autoclaving, these two solutions should be mixed together aseptically and thoroughly. Four ml of a filter-sterilized, 10 mg/ml solution of the required amino acid is then added and mixed in well. The medium is then poured into plates.

Use This medium is used to demonstrate growth-factor requirements of organisms. It will support the growth of *E. coli* strains auxotrophic for the amino acid(s) added.

Mineral Salts + Glucose + Streptomycin

K$_2$HPO$_4$	7.0 g
KH$_2$PO$_4$	2.0
(NH$_4$)$_2$SO$_4$	1.0
MgSO$_4$	0.1
Glucose	5.0
Agar	15.0
Distilled H$_2$O	1000 ml
pH 7.0 ± 0.2	
Streptomycin	50 mg
Leucine	40 mg
Thiamine	1 mg

Preparation This medium should be prepared in separate solutions for best results. Solution A contains the first three ingredients in 500 ml distilled H$_2$O and is adjusted to pH 7.0 ± 0.2. Solution B contains the MgSO$_4$, glucose and agar in 500 ml of distilled H$_2$O. These solutions are autoclaved, then mixed aseptically and thoroughly. To this mixture, the following amounts of filter-sterilized streptomycin, leucine, and thiamine are added: 5 ml of a 10 mg/ml solution of streptomycin; 4 ml of a 10 mg/ml solution of leucine; and 1 ml of a 1 mg/ml solution of thiamine (vitamin B$_1$). These last two ingredients are added because the ATCC F-strain used in the genetics exercise is auxotrophic for these growth factors.

Use This medium is used to detect threonine-positive, streptomycin-resistant recombinants in the conjugation of *E. coli Hfr* (prototroph, Str^s) and *E. coli* F$^-$(Thr$^-$, Leu$^-$, Str^R, B$_1^-$).

Mineral Salts + Yeast extract + Glucose

KH$_2$PO$_4$	2.0 g
K$_2$HPO$_4$	7.0
(NH$_4$)$_2$SO$_4$	1.0
MgSO$_4$	0.1
Yeast extract	0.1
Glucose	5.0
Agar	15.0
Distilled H$_2$O	1000 ml
pH 7.0 ± 0.2	

Preparation Do not overheat. Medium should not be brown.

Use This medium is used to determine the utilization by an organism of glucose as a carbon and energy source. The small amount of yeast extract provides vitamins and amino acids, but is not present in sufficient quantity to support substantial visible growth if used as the sole carbon and energy source.

Mineral Salts + Yeast extract + Lactose

KH$_2$PO$_4$	2.0 g
K$_2$HPO$_4$	7.0
(NH$_4$)$_2$SO$_4$	1.0
MgSO$_4$	0.1
Yeast extract	0.1
Lactose	5.0
Agar	15.0
Distilled H$_2$O	1000 ml
pH 7.0 ± 0.2	

Preparation Do not overheat. Medium should not be brown.

Use This medium is used to determine the utilization by an organism of lactose as a carbon and energy source. The small amount of yeast extract provides vitamins and amino acids, but is not present in sufficient quantity to support substantial visible growth if used as the sole carbon and energy source.

Mineral Salts + Yeast extract + Sucrose

KH$_2$PO$_4$	2.0 g
K$_2$HPO$_4$	7.0
(NH$_4$)$_2$SO$_4$	1.0
MgSO$_4$	0.1
Yeast extract	0.1
Sucrose	5.0
Agar	15.0
Distilled H$_2$O	1000 ml
pH 7.0 ± 0.2	

Preparation Do not overheat. Medium should not be brown.

Use This medium is used to determine the utilization by an organism of sucrose as a carbon and energy source. The small amount of yeast extract provides vitamins and amino acids, but is not present in sufficient quantity to support substantial visible growth

if used as the sole carbon and energy source.

Motility Medium

Glucose	2.0 g
Peptone	3.0
Yeast extract	2.0
Agar	5.0
Distilled H$_2$O	1000 ml
pH 7.0 ± 0.1	

Preparation This medium should be prepared in clear, unscratched tubes to allow observation of turbidity in and away from the line of inoculation.

Use This medium is used to detect motility of a culture as indicated by turbidity away from the line of inoculation. The low level of agar allows motile organisms to move through the medium, but prevents nonmotile organisms from moving simply due to convection currents. Organisms which are obligately aerobic will often not grow below the surface in this medium and alternative motility tests must be used.

Neisseria Fermentation Medium

Beef extract	1.0 g
Proteose peptone #3	10.0
NaCl	5.0
Phenol red	0.018
Agar	8.0
Cornstarch	1.0
Carbohydrate	5.0
glucose (or sucrose, as required)	
Distilled H$_2$O	1000 ml
pH 7.4	

Preparation This medium must be steamed to swell the starch before dispensing into tubes. The color of the medium should be light orange, not yellow or red.

Use This medium is used to detect the production of acid from the fermentation of carbohydrates by *Neisseria*. Since the *Neisseria* do not produce much acid, it is important to compare the color of inoculated tubes with uninoculated controls to detect the yellow, positive result. The cornstarch in the medium is added to counteract any toxicity of peptones for *Neisseria*. (This starch could be fermented by other organisms, so anomalous positive reactions should be rechecked for culture purity.)

Nitrogen-free Agar

K$_2$HPO$_4$	0.8 g
KH$_2$PO$_4$	0.2
MgSO$_4$	0.2
CaSO$_4$	0.1
FeSO$_4$·7H$_2$O	0.003
MoO$_3$	0.001
Sucrose	5.0
Agar	15.0
Distilled H$_2$O	1000 ml
pH 7.6 ± 0.1	

Preparation This medium should be prepared as two separate solutions for best results. Solution A contains the first two ingredients in 500 ml of distilled H_2O and is adjusted to a pH of 7.6 ± 0.1. Solution B contains the rest of the ingredients in 500 ml of distilled H_2O. After autoclaving, the two solutions are mixed together aseptically and thoroughly.

Use This medium is used for the isolation of free-living nitrogen-fixing organisms.

Nitrogen-free Broth

K_2HPO_4	0.8 g
KH_2PO_4	0.2
$MgSO_4$	0.2
$CaSO_4$	0.1
$FeSO_4 \cdot 7H_2O$	0.003
MoO_3	0.001
Sucrose	5.0
Distilled H_2O	1000 ml
pH 7.6 ± 0.1	

Preparation For enrichment purposes, this medium need not be sterilized. If it is to be sterilized, two solutions should be prepared. The first two ingredients should be prepared in 500 ml of distilled H_2O, and the pH adjusted. The other ingredients should be added to another 500 ml of distilled H_2O. The separate solutions should be autoclaved, then mixed aseptically.

Use This medium is used for the enrichment and cultivation of free-living nitrogen-fixing organisms. Notice the addition of extra iron and molybdenum which are required as cofactors for the nitrogenase enzyme.

Nutrient Agar

Beef extract	5.0 g
Peptone	3.0
Agar	15.0
Distilled H_2O	1000 ml
pH 6.8–7.0	

Preparation Also available commercially in dehydrated form.

Use For isolation and cultivation of organisms that are not nutritionally fastidious.

Nutrient Agar + Streptomycin

Beef extract	5.0 g
Peptone	3.0
Agar	15.0
Distilled H_2O	1000 ml
pH 6.8–7.0	

Preparation Nutrient agar is available commercially in dehydrated form. The antibiotic(s) should be prepared in solution, *filter-sterilized* and added aseptically in appropriate amounts after the nutrient agar has been autoclaved and cooled to 50–65°. Add 1 ml/liter of a 10 mg/ml solution of streptomycin to achieve a final concentration of 10 μg/ml in the agar. Add 10 ml to achieve 100 μg/ml in the agar. Mix well, then pour into plates.

Use This medium is used to isolate mutants resistant to the antibiotic used.

Nutrient Agar + Sucrose

Beef extract	5.0 g
Peptone	3.0
Sucrose: 5%	50
or	
Sucrose: 10%	100
or	
Sucrose: 20%	200
Agar	15.0
Distilled H_2O	1000 ml
pH 6.8–7.0	

Preparation Nutrient agar is also available commercially in dehydrated form. This can be prepared and the appropriate amount of sucrose added to it before autoclaving. Do not overheat.

Use For the study of the effects of decreasing water activity (increasing osmotic potential) on bacterial growth.

Nutrient Broth

Beef extract	5.0 g
Peptone	3.0
Distilled H_2O	1000 ml
pH 6.8–7.0	

Preparation Also available commercially in dehydrated form.

Use For cultivation of organisms that are not nutritionally fastidious.

Nutrient Broth + 1% Glucose

Beef extract	5.0 g
Peptone	3.0
Glucose	10.0
Distilled H_2O	1000 ml
pH 6.8–7.0	

Preparation Nutrient broth to which the glucose can be added is also available commercially in dehydrated form.

Use This medium can be used for the cultivation of nonfastidious organisms.

Nutrient Broth + 1% Lactose

Beef extract	5.0 g
Peptone	3.0

Lactose	10.0
Distilled H$_2$O	1000 ml
pH 6.8–7.0	

Preparation   Nutrient broth to which the lactose can be added is available commercially in dehydrated form. Do not overheat.

Use   This medium can be used for the cultivation of nonfastidious organisms. In addition, the lactose will induce the synthesis of the enzyme β-galactosidase. This can be detected by adding the chromogenic substrate, ONPG, after the cells have been treated with toluene to make their cell membranes permeable (see β-galactosidase Assay, Student Supplement 1).

Peptone-Succinate Broth

(NH$_4$)$_2$SO$_4$	1.0 g
MgSO$_4$·7H$_2$O	1.0
MnSO$_4$·H$_2$O	0.002
FeCl$_3$·6H$_2$O	0.002
Sodium succinate	2.0
Peptone	5.0
Distilled H$_2$O	1000 ml
ph 7.0 ± 0.1	

Preparation   If a solid medium is needed, add 15g of agar/liter.

Use   For the cultivation of *Spirillum*.

Photosynthetic Mineral Salts Agar

KH$_2$PO$_4$	0.33 g
MgSO$_4$·7H$_2$O	0.33
NaCl	0.33
NH$_4$Cl	0.5
CaCl$_2$·2H$_2$O	0.05
Na succinate	1.0
Yeast extract	0.02
Agar	18.0
Distilled H$_2$O	990 ml
pH 6.9 ± 0.1	

After autoclaving, add the following separately filter-sterilized solutions:

Trace salts	1.0 ml
(see below)	
FeSO$_4$ solution	0.5 ml
(200 mg FeSO$_4$·7H$_2$O per liter)	
NaHCO$_3$ solution	10.0 ml
(10g NaHCO$_3$/100ml)	

Trace salts

ZnSO$_4$·7H$_2$O	10 mg
MnCl$_2$·4H$_2$O	3
H$_3$BO$_3$	30
CoCl$_2$·6H$_2$O	20
CuCl$_2$·2H$_2$O	1
NiCl$_2$·6H$_2$O	2
NaMoO$_4$	3
Distilled H$_2$O	1000 ml

Preparation   For enrichment purposes, this medium need not be sterilized. The trace salts solution can be stored for long periods of time, but the FeSO$_4$ and NaHCO$_3$ solutions should be prepared fresh.

Use   This medium is used for the isolation and cultivation of photosynthetic bacteria of the family Rhodospirillaceae. It must be incubated anaerobically in the light.

Photosynthetic Mineral Salts Broth

KH$_2$PO$_4$	0.33 g
MgSO$_4$·7H$_2$O	0.33
NaCl	0.33
NH$_4$Cl	0.5
CaCl$_2$·2H$_2$O	0.05
Na succinate	1.0
Yeast extract	0.02
Distilled H$_2$O	990 ml
pH 6.9 ± 0.1	

After autoclaving, add the following separately filter-sterilized solutions:

Trace salts	1.0 ml
(see below)	
FeSO$_4$ solution	0.5 ml
(200 mg FeSO$_4$·7H$_2$O per liter)	
NaHCO$_3$ solution	10.0 ml
(10 g NaHCO$_3$/100 ml)	

Trace salts

ZnSO$_4$·7H$_2$O	10 mg
MnCl$_2$·4H$_2$O	3
H$_3$BO$_3$	30
CoCl$_2$·6H$_2$O	20
CuCl$_2$·2H$_2$O	1
NiCl$_2$·6H$_2$O	2
NaMoO$_4$	3
Distilled H$_2$O	1000 ml

Preparation   For enrichment purposes, this medium need not be sterilized. The trace salts solution can be stored for long periods of time, but the FeSO$_4$ and NaHCO$_3$ solutions should be prepared fresh.

Use   This medium is used for the enrichment of photosynthetic bacteria in the family Rhodospirillaceae. The medium must be incubated anaerobically in the light. If the containers are not sealed, aerobic organisms will grow, utilizing the succinate, a nonfermentable carbohydrate. The best light to use for incubation is either natural light (north window, to avoid excessive heating on sunny days) or tungsten lamps, at 25 cm distance if using a 25 watt bulb.

Plate Count Agar (Standard Methods Agar)

Tryptone or trypticase	5.0 g
Yeast extract	2.5
Glucose	1.0
Agar	15.0
Distilled H$_2$O	1000 ml

Preparation   Available commercially in dehydrated form.

Use   This medium is used for the cultivation, isolation, and enumeration of nonfastidious organisms.

Potassium Phosphate Buffer

Preparation   To prepare potassium phosphate buffers, prepare 2 separate solutions of $1M$ K_2HPO_4 (174.2 g/l) and $1M$ KH_2PO_4 (136.1 g/l). Using a pH meter or pH paper, mix the two solutions until the proper pH is obtained. This will give a $1M$ potassium phosphate buffer. Dilute as necessary(1/10 to give $10^{-1}M$, 1/100 to give $10^{-2}M$ and 1/1000 to give $10^{-3}M$).

Use   This is used as a buffer base for many media. Also used as a buffer for washing leaves.

Pseudomonas F Agar

Tryptone	10.0 g
Proteose peptone #3	10.0
Glycerol	10 ml
K_2HPO_4	1.5 g
$MgSO_4$	1.5
Agar	15.0
Distilled H_2O	990 ml

Preparation   Available commercially in dehydrated form, to which the glycerol must be added.

Use   This medium is used to detect the production of the ironophore, fluorescein. The low iron levels induce the production of high levels of flourescein, which can be detected using an ultraviolet light source.

Rhizobium Isolation Agar

Mannitol	10.0 g
Yeast extract	1.0
$MgSO_4 \cdot 7H_2O$	0.2
NaCl	0.2
K_2HPO_4	0.5
$FeCl_3$	0.005
Agar	15.0
Distilled H_2O	1000 ml

Use   This medium is used to isolate and cultivate *Rhizobium*. *Rhizobium* colonies can be easily distinguished by their clear (sometimes whitish), slimy colonies. Using properly surface-sterilized nodules, *Rhizobium* should be almost the only organism on the plate. This medium is *not* selective, so proper surface-sterilization of nodules is critical.

Sabouraud Dextrose Agar

Neopeptone	10.0 g
Dextrose (glucose)	40.0
Agar	15.0

Preparation   Available commercially in dehydrated form. Final pH will be about 5.6.

Use   For the cultivation of fungi.

Starch Agar

Beef extract	5.0 g
Peptone	3.0
Starch (soluble)	4.0
Agar	15.0
Distilled H_2O	1000 ml
pH 6.8–7.0	

Preparation   Nutrient Agar, to which the starch can be added before autoclaving, is available commercially in dehydrated form.

Use   This medium is used to detect the production of the extracellular enzyme amylase. This enzyme hydrolyzes the starch in the medium surrounding the colony, resulting in a clear halo around the colony after addition of iodine solution. Unhydrolyzed starch will react with the iodine to give a dark blue color (see Amylase Test, Student Supplement 1).

Streptomyces Agar

Yeast extract	1.0 g
Tryptone	10.0
Soluble starch	10.0
K_2HPO_4	0.5
Agar	15.0
Distilled H_2O	1000 ml
pH 7.3 ± 0.1	

Use   This medium is used for the isolation and cultivation of *Streptomyces*. The medium promotes early sporulation. It also supports the growth of a wide range of other organisms that might be used in the testing of antibiotic production by members of the *Streptomyces* group.

Sucrose Agar Deeps

Tryptone	10.0 g
Yeast extract	3.0
Bromcresol purple	0.02
Sucrose	10.0
Agar	12.0
Distilled H_2O	1000 ml
pH 7.4 ± 0.1	

Preparation   As mentioned for Glucose Agar Deeps, the bromcresol purple dye can be added as a solid or as a solution. Tubes must be melted and cooled to 45–50° for inoculation, then cooled to solidify for incubation.

Use   This medium is used to detect acid and gas production in the fermentation of sucrose. Acid production will lower the pH of the medium, resulting in a change of the pH indicator, bromcresol purple, from purple to yellow. Gas will be evident by the break-up of the agar in the tube.

Sucrose Durham Tube

Tryptone	10.0
Yeast extract	3.0

Bromcresol purple	0.02
Sucrose	10.0
Distilled H$_2$O	100 ml
pH 7.4±0.1	

Preparation As mentioned in Glucose Agar Deeps, the dye can be added from a stock solution rather than a solid. The medium should be dispensed into tubes which contain a smaller inverted tube. Fill the larger tube about 3/4 full of medium. Autoclave. As the tubes cool after autoclaving, the small inverted tube will fill with the broth.

Use This medium is used to detect the production of acid and gas in the fermentation of sucrose. Acid production will lower the pH of the medium, resulting in a change of the pH indicator, bromcresol purple, from purple to yellow. Gas will be evident as a bubble in the inverted tube. Tubes with negative results for gas production should be tapped lightly on the bottom and observed for very small gas bubbles rising to the surface. Those tubes showing these small bubbles should be recorded as weakly positive for gas production.

Sucrose Gelatin Agar

Tryptone	10.0 g
Yeast extract	5.0
NaCl	5.0
K$_2$HPO$_4$	1.0
Glucose	1.0
Sucrose	50.0
Gelatin	50.0
Agar	20.0
Distilled H$_2$O	1000 ml

Preparation Available commercially in dehydrated form. Steam to completely dissolve the gelatin *before autoclaving*. Handle this medium with care when first removing it from the autoclave, as it is often superheated even after a proper slow exhaust cycle and might spurt out of the flask opening.

Use This medium is used to detect the formation of dextran or levan from the utilization of sucrose.

Sucrose Gelatin Agar + NaN$_3$

Ingredients as in Sucrose Gelatin Agar to which 0.2g of NaN$_3$ is added before autoclaving.

Preparation Prepare as in procedure for Sucrose Gelatin Agar.

Use This medium is used for the isolation of lactic acid bacteria capable of producing levan or dextran from sucrose. The NaN$_3$ inhibits organisms with a respiratory system.

Thioglycollate Agar (for tubes)

Yeast extract	5.0 g
Casitone	15.0

Glucose	5.0
NaCl	2.5
Cystine	0.75
Thioglycollic acid	0.3 ml
Agar	3.5 g
Resazurin	0.001
Distilled H$_2$O	1000 ml
pH 7.1±0.1	

Preparation Fluid thioglycollate broth is available commercially in dehydrated form. The agar content of the broth is 0.75 g/l, so an additional 2.8g of agar should be added. Do not store under refrigeration or oxygen will diffuse deep into the tubes. The medium must be melted and cooled to 45–50° for inoculation.

Use For use in the determination of oxygen requirements for growth. Its advantage over Fluid Thioglycollate Broth for introductory students is that surface growth does not descend during the incubation period which could lead to false results. The bottom of the tube of medium is kept anaerobic by cystine and thioglycollic acid, which chemically react with and tie up any oxygen which diffuses in. Any unreacted oxygen in the tube will be indicated by the resazurin dye, which turns pink in the presence of oxygen. It is common for the top centimeter or so to be pink.

Top Agar

Peptone	0.6 g
Yeast extract	3.0
NaCl	2.0
CaCl$_2$	0.1
Agar	7.0
Distilled H$_2$O	1000 ml

Preparation Dissolve the medium ingredients using heat (100°) then dispense into tubes, 3 ml/tube. Sterilize in an autoclave. Before use, melt the agar and cool to 45–50° for inoculation.

Use This medium is designed as an overlay for bacteriophage detection. It is inoculated with bacteria and possibly bacteriophage while liquid (45–50°), then poured onto Bottom Agar. The plate should be immediately tilted to distribute the overlay. After incubation, a bacterial lawn will be evident as turbidity, whereas bacterial lysis will be indicated by a clear area. The low level of agar allows for diffusion of the bacteriophage particles and the CaCl$_2$ is required for their attachment to the bacterial host. This medium can also be used as an overlay for other tests requiring bacterial lawns (for instance, antibiotic or disinfectant tests).

Tryptone Broth

Tryptone	10 g
Distilled H$_2$O	1000 ml
pH 7.2±0.1	

Use This medium contains high levels of the amino acid tryptophan and is used to detect the production of the enzyme tryptophanase. If the organism produces the enzyme, the tryptophan will be degraded to pyruvic acid, NH_4^+ and indole. Indole can be detected by a red ring on the surface of the broth after the addition of Kovac's reagent (see Indole Test, Student Supplement 1).

Tryptone Yeast Extract Agar

Tryptone	10.0 g
Yeast extract	3.0
Distilled H_2O	1000 ml

Use For the cultivation of both *Streptomyces* and test organisms for a test of antibiotic production.

Urea Broth

Urea	20.0 g
KH_2PO_4	9.1
K_2HPO_4	9.5
Yeast extract	0.1
Phenol red	0.01
Distilled H_2O	1000 ml
pH 6.8	

Preparation This medium must be filter sterilized as heat destroys the urea. After filter sterilization, it can be aseptically distributed in 3–4 ml amounts into sterile tubes. The medium should not be red.

Use This medium is used to detect the production of the enzyme urease, which hydrolyzes urea to produce NH_4^+. This alkaline product is detected by the change of the pH indicator, phenol red, to a red-violet color. This reaction can occur in as little as 3–4 hours, depending on the organism being tested.

Vogel-Johnson Agar

Peptone	10.0 g
Yeast extract	5.0
Mannitol	10.0
K_2HPO_4	5.0
LiCl	5.0
Glycine	10.0
Agar	16.0
Phenol red	0.025
Distilled H_2O	1000 ml
pH 7.2 ± 0.1	
After autoclaving, add filter-sterilized solution of:	
Potassium tellurite	20.0 ml
(1 g/100 ml H_2O)	

Preparation Available commercially in dehydrated form as Vogel-Johnson Agar Base, to which the filter-sterilized potassium tellurite must be added.

Use This medium is used to isolate *Staphylococcus aureus*. The lithium and tellurite select against other organisms. Coagulase-positive, mannitol-fermenting *S. aureus* will appear as large black colonies (due to reduction of tellurite to elemental tellurium) surrounded by a yellow halo (due to the production of acids from mannitol).

Yeast Extract Glucose Agar

Peptone	3.0 g
Yeast extract	2.0
Glucose	10.0
Agar	15.0
Distilled H_2O	1000 ml

Preparation The pH of this medium can be adjusted to 4.0, 5.0, and 6.0, with 6N HCl, to 9.0 with 6N KOH or NaOH and to 7.0 with the acid or base as needed.

Use For the determination of pH ranges for growth.

Reagents

Acid-fast Stain

Carbol Fuchsin

Solution A:

Basic fuchsin	3	g
Ethyl alcohol	100	ml

Solution B:

Phenol, melted crystals	50	ml
Distilled H_2O	950	ml

Preparation In a hood, prepare Solutions A and B.

The phenol crystals in Solution B will melt at 60°C. It is easiest to do this by putting the manufacturer's bottle directly in a 60°C H_2O bath until the crystals melt, then pipette the correct amount of phenol. CAUTION! Phenol is a highly toxic and caustic chemical. It is absorbed through the skin and can cause death even after a short exposure. Wear goggles, gloves and chemical (rubberized) apron when handling. Do not mouth pipette. Clean up all spills immediately. Do not attempt to weigh out phenol crystals as the chopping required to break up the mass of crystals might result in generalized contamination of the lab area by phenol. Mix Solutions A and B together and let the mixture stand for 2–3 days before use.

Acid-Alcohol

95% Ethanol	1000 ml
Concentrated HCl	30 ml

Preparation   Mix the two liquids together as listed, adding the acid to the alcohol. Work in a hood as concentrated HCl is a volatile, caustic chemical.

Methylene Blue

Use the same methylene blue as for the simple stain (see Methylene Blue).

Flagella Stain (Ryu's)

Solution A

5% Aqueous phenol solution	10 ml
Tannic acid	2 g
Saturated aqueous potassium aluminum sulfate	10 ml

Solution B

Saturated alcoholic crystal violet solution (approximately 1.2 g crystal violet in 10 ml 95% ethanol).	

Preparation   To prepare the 5% aqueous phenol solution, melt the phenol crystals as indicated for the preparation of the Carbol Fuchsin in the Acid-fast Stain. CAUTION! Phenol is a highly toxic and caustic chemical. Observe all the precautions outlined for the preparation of Carbol Fuchsin. After the crystals have melted, add 5 ml of phenol to 100 ml of distilled H_2O. Mix well and dispense before the solvents separate. When preparing Solution B, be careful weighing out the crystal violet to avoid spreading the crystals around the laboratory area. Crystal violet spills can be cleaned up with 95% ethanol.
Prior to use, mix 10 volumes of well-mixed Solution A with 1 volume of Solution B. Filter if a precipitate forms. Although Solutions A and B are stable for long periods under refrigeration, the mixture is only effective on the day it is prepared.

Gram's Reagents

Gram's Crystal Violet

Solution A

Crystal violet	20 g
95% Ethanol	200 ml

Solution B

Ammonium oxalate	8 g
Distilled H_2O	800 ml

Preparation   Prepare Solutions A and B, mix the two solutions together well and let stand overnight. Filter with 2 layers of filter paper (for example, Whatman-Reeve Angel Grade 202) to remove undissolved crystals. Be careful when weighing out the crystal violet to avoid spreading the crystals around the laboratory area. Crystal violet spills can be cleaned up with 95% ethanol.

Gram's Iodine

Iodine crystals	3.3 g
KI	6.7
Distilled H_2O	1000 ml

Preparation   Grind the iodine and KI crystals together using a mortar and pestle. Slowly add the water while grinding. The final solution should be a deep golden color. It will turn to light yellow with age and should be discarded.

Gram's Alcohol-Acetone

Acetone	20 ml
95% Ethanol	980 ml

Preparation   Both of these solvents are flammable and harmful when ingested. In addition, acetone can be absorbed through the skin, so precautions should be taken to avoid spillage. Do not mouth pipette this solution. Do not use it to remove crystal violet from the skin, instead use 95% ethanol.

Gram's Safranin

Safranin	25 g
95% Ethanol	100 ml
Distilled H_2O	900 ml

Preparation   As with all dyes, this should be weighed out carefully to avoid spreading dye crystals around the laboratory area.

Hydrogen Peroxide (5% H_2O_2)

30% H_2O_2	40 ml
Distilled H_2O	200 ml

Preparation   30% H_2O_2 is available commercially. It is a strong oxidant and should be handled with care, not mouth pipetted, and refrigerated when stored. The 5% H_2O_2 solution used in the catalase test should be discarded after one week as it decomposes to O_2 and H_2O spontaneously.

Kovac's Reagent

para-Dimethyl-amino-benzaldehyde	5.0 g
Butanol	75.0 ml
Concentrated HCl	25.0 ml

Preparation   This reagent should be mixed in a hood. Wear eye protection. Do not mouth pipette the butanol or the concentrated HCl. Clean up spills immediately.

Methylene Blue

Methylene blue	1 g
Distilled H_2O	100 ml

Preparation   As with all dyes, this should be weighed out carefully to avoid spreading dye crystals around the laboratory area.

Methyl-red Reagent

Methyl red	0.2 g
95% Ethanol	600 ml
Distilled H$_2$O	400 ml

Preparation   Weigh out the dye carefully to avoid spreading crystals around the laboratory area.

ONPG (ortho-nitrophenyl-β-D-galactoside)

ONPG	2.0 g
10^{-3}M Potassium phosphate buffer, pH 7.6	500 ml

Preparation   ONPG is available from a variety of vendors such as Sigma and Calbiochem (see vendor list).
The phosphate buffer is prepared as described in the media section. This reagent should be clear and refrigerated until use. Discard if it turns light yellow.

Oxidase Reagent

Tetra-methyl-para-phenylenediamine HCl	0.5 g
Distilled H$_2$O	100 ml

Preparation   Mix rapidly. The reagent should be clear or light pink, not purple. Discard after one day.

Spore Stain

Malachite Green Solution

Malachite green	50.0 g
Distilled H$_2$O	1000 ml

Preparation   Weigh the dye carefully to avoid spreading crystals around the laboratory area. Mix well, let sit overnight, then filter through 2 layers of filter paper (for example, Whatman-Reeve Angel Grade 202) to remove undissolved crystals.

Safranin Solution

Safranin	5.0 g
Distilled H$_2$O	1000 ml

Preparation   Weigh the dye carefully to avoid spreading crystals around the laboratory area.

Toluene in Acetone

Toluene	5 ml
Acetone	95 ml

Preparation   Both toluene and acetone are harmful and flammable solvents. The solution should be mixed in a hood, taking care not to spill either on the hands. This solution should not be mouth pipetted!

Voges-Proskauer Reagents

α-naphthol

α-naphthol	5.0 g
95% Ethanol	100 ml

Preparation   The α-naphthol is expensive so the volume of reagent prepared should be as near as possible to the required amount.

40% KOH

KOH	40 g
Distilled H$_2$O	100 ml

Preparation   This reagent will become hot when mixed so prepare in a proper container (Pyrex, Kimax). This reagent is also caustic. Clean up spills immediately.

Notes

Dilution Theory and Problems

Microorganisms are often enumerated (counted) in the laboratory using such methods as the viable plate count where a dilution of the sample is plated onto (or into) an agar medium. After incubation, plates with 30–300 colonies per standard-sized plate are counted. At this colony number, the number counted is high enough to have statistical accuracy, yet low enough to avoid nutrient competition among the developing colonies. Each of the colonies is presumed to have arisen from only one cell, although this may not be true if pairs, chains, or groups of cells are not completely broken apart before plating.* It is possible, but unlikely, for an original (undiluted) sample of microorganisms which is to be counted to have 30–300 cells/ml so that a pour plate using a 1 ml volume from the sample will give good results.† More likely, a sample will have greater numbers of cells/ml; sometimes, as in the case of unpolluted water samples, the sample will have less. In either case, the sample must be manipulated so that it contains a number of cells in the correct range for plating. If the cell numbers are too high, the sample is diluted; if too low, the sample is concentrated. Dilutions are performed by careful, aseptic pipetting of a known volume of sample into a known volume of a sterile buffer or sterile water. This is mixed well and can be used for plating and/or further dilution. If the number of cells/ml is unknown, then a wide range of dilutions are usually prepared and plated.

Concentration of samples is usually performed by filtration through a filter with pores small enough to retain the microorganisms. This way, the volume of the original sample can be reduced, so that the number of cells/ml increases to the "countable" range. Since the concentration procedure is not used in this manual, this supplement will focus on dilution theory.

In order to make the calculation of the number of cells/ml in the original sample less formidable, dilutions are designed to be easy to handle mathematically. The most common dilutions are 1/10, 1/100, and 1/1000. Looking first at the 1/10 (or 10^{-1}) dilution‡, it can be achieved by mixing 1 ml of sample with 9 ml of sterile dilution buffer. This gives the fraction:

$$\frac{1 \text{ ml of sample}}{1 \text{ ml of sample} + 9 \text{ ml of buffer}} = \frac{1 \text{ ml}}{1 \text{ ml} + 9 \text{ ml}} = \frac{1 \text{ ml}}{10 \text{ ml}} = \frac{1}{10} = 10^{-1}$$

Some alternative methods of obtaining a 1/10 dilution are:

$$\frac{10 \text{ ml of sample}}{10 \text{ ml of sample} + 90 \text{ ml of buffer}} = \frac{10 \text{ ml}}{10 \text{ ml} + 90 \text{ ml}} = \frac{10 \text{ ml}}{100 \text{ ml}} = \frac{1}{10}$$

$$\frac{0.1 \text{ ml of sample}}{0.1 \text{ ml of sample} + 0.9 \text{ ml of buffer}} = \frac{0.1}{0.1 + 0.9} = \frac{0.1}{1.0} = \frac{1}{10}$$

$$\frac{5 \text{ ml of sample}}{5 \text{ ml of sample} + 45 \text{ ml of buffer}} = \frac{5}{5 + 45} = \frac{5}{50} = \frac{1}{10}$$

What if the sample is not liquid, such as in food samples? Since 1 ml of water weighs 1 gram (under specific standard conditions), a gram of any sample is arbitrarily assigned the volume of 1 ml. So 1 gram of pepper + 9 ml of buffer = a 1/10 dilution of the pepper. It is recognized that 1 gram of pepper might not *actually* have the volume of 1 ml, but if the arbitrary assignment is recognized as standard, it can be used conveniently and reproducibly.

How is a 1/100 dilution obtained? Usually, 1 ml of sample + 99 ml buffer or 0.1 ml of sample + 9.9 ml of buffer are used to give this dilution. A 1/1000 dilution can be obtained by adding 0.1 ml of sample to 99.9 ml of buffer.

Once the dilution is made, an aliquot can be plated on/in an agar medium. After incubation, the colonies can be counted. How do the colonies on the plate relate to the number of cells (colony-forming units) in the original sample? Try a problem:

- One ml of a sample was mixed with 99 ml of buffer. One ml of this was plated (using the pour plate method) in nutrient agar. After incubation, 241 colonies were present on the plate. How many colony-forming units were present per ml of the *original sample*?

The dilution used was $1/(1 + 99) = 1/100 = 10^{-2}$. One ml of the dilution contained 241 colony-forming units. How much did one ml of the original sample contain? Obviously, *more than* 241 colony-forming units! To arrive at the correct number, *either* divide the colony-forming units/ml of the dilution by the dilution (241 colony-forming units/ml $\div 10^{-2} = 241 \times 10^2$ $= 2.41 \times 10^4$ colony-forming units per ml of original sample) or multiply by the *dilution factor*. The dilution factor is defined as the *inverse of the dilution*, so in this case would be $1/10^{-2} = 10^2$. Multiplying by the dilution factor: 241 colony-forming units/ml $\times 10^2 = 241 \times 10^2 = 2.41 \times 10^4$ colony-forming units/ml of original sample. Either way, the answer obtained is the same. Try another:

- One ml of a sample was mixed with 99 ml of buffer. One-tenth of a ml of this was plated on nutrient agar. After incubation, 142 colonies were present on the plate. How many colony-forming units were present per ml of the original sample?

In this case, the mathematics of the dilutions is the same, but the number of colonies counted on the plate represent the number of colony-forming units in *only 0.1 ml* of the dilution plated. Therefore, the number of colony-forming units per ml of dilution is 142 colonies formed $\div$ 0.1 ml plated, or $142 \times 10 = 1420$ colony-forming units/ml of dilution. Dividing this number by the 10^{-2} dilution results in a final answer of 1420×10^2 or 1.42×10^5 colony-forming units/ml of original sample. Try another:

- One ml of a sample was mixed with 99 ml of buffer. One ml of this was added to a sterile petri plate and mixed with 25 ml of molten agar (cooled to 45°). The mixture was allowed to solidify undisturbed. After incubation, 241 colonies were present on the plate. How many colony-forming units were present per ml of the original sample?

The answer to this problem is *identical* to the answer to the first problem! The number of ml of agar used is not relevant because the number of organisms present depends only on the number of ml of the dilution which were added to the agar. Whether 25 ml or 35 ml of agar were used, the number of organisms added to the agar remains constant, so the colony count will be identical.

What if the sample requires more dilution? For example, what if the sample to be counted is a culture of *E. coli* grown overnight in a rich medium? The number of cells/ml will be around 10^9 (1,000,000,000). Obviously, a single dilution would not be enough, so *successive dilutions* must be performed. An easy example of a successive dilution is making a 1/100 dilution using only two tubes of 9.0 ml of buffer. Add 1 ml of sample to the first tube and mix well. This is a 1/10 dilution. Now remove 1 ml from this dilution and add it to the second buffer tube. This is also a 1/10 dilution, *but not of the original sample*. Instead, this is a 1/10 dilution of a 1/10 dilution. To arrive at the final dilution, multiply the second dilution by the first dilution: $1/10 \times 1/10 = 1/100$! Another 1/10 dilution of this second dilution will result in a 1/1000 final dilution. Try a problem:

- An overnight culture of *E. coli* is used as a sample. One ml of this culture is added to a bottle containing 99 ml of buffer. This dilution is mixed well (as all dilutions are!), and one ml of this is mixed in 9 ml of buffer. This second dilution is diluted by three successive 1/10 dilutions. The last (fifth) dilution is used for plating, and one-tenth ml is plated on nutrient agar. After incubation, 56 colonies are counted on the plate. How many colony-forming units were present per ml of *E. coli* culture?

To solve this problem, try writing out the procedure. Now multiply the successive dilutions together (here is where scientific notation comes in handy, since it is easy to add the exponents without "losing a zero").

$$\frac{1}{1 + 99} \times \frac{1}{1 + 9} \times \frac{1}{10} \times \frac{1}{10} \times \frac{1}{10} = \frac{1}{1,000,000}$$

OR

$$10^{-2} \times 10^{-1} \times 10^{-1} \times 10^{-1} \times 10^{-1} = 10^{-6}$$

The number of colony-forming units per ml of the dilution can then be divided by the final dilution:

$$\frac{56 \text{ colonies}}{0.1 \text{ ml plated}} \div 10^{-6} = 560 \text{ colony-forming units/ml} \times 10^6$$

$$= 5.6 \times 10^8 \text{ colony-forming units/ml}$$

Note that the 0.1 ml plated acts as another 1/10 dilution. Some microbiologists treat it as such in their computations rather than dividing the colony count by the number of milliliters plated, as done here. Whichever way it is done, the answer will be the same.

The only way to understand dilution theory well is to practice it, so you should work the problems given in this supplement. The answers to the problems are given at the end. Before you start, though, remember a few basic rules of dilution theory:

- Since dilutions result in *physically* lowering the number of cells per ml in a solution, the calculations must *mathematically* raise the number of cells per ml. Therefore, remember that the final answer should be *larger* than the original colony count, not smaller. It is not possible to have negative exponents in the final number of colony-forming units/ml.
- The number of ml of agar into which a sample is added is irrelevant, unless of course, it is further diluted.
- The number of ml plated is relevant.
- The dilution is determined by dividing the number of ml of sample added to the dilution buffer by the *total* of the ml of sample plus the ml of buffer.
- Successive dilutions are multiplied to find the total dilution.
- Plates with 30–300 colonies on them should be used for the greatest accuracy in counting.
- The number of colony-forming units/ml of original suspension is calculated by either dividing the number of colony-forming units/ml of the dilution which gave 30–300 colonies per plate by that dilution *or* by multiplying the number of colony-forming units/ml of the dilution which gave 30–300 colonies per plate by the inverse of that dilution (i.e. the dilution factor).
- If more than one plate of a dilution has been prepared and is counted, the colony counts (of that dilution only) should be averaged.

Questions

1. One ml of a water sample is added to 9 ml of sterile water. This is mixed well and further diluted by 4 successive 1/10 dilutions. One-tenth of a ml of each dilution is spread on a plate of nutrient agar. After incubation, the following colony counts were obtained:

Dilution used for plating	Amount plated	Colony counts after incubation
first 1/10	0.1 ml	too many to count
second 1/10	0.1 ml	730
third 1/10	0.1 ml	67
fourth 1/10	0.1 ml	5
fifth 1/10	0.1 ml	0

What was the number of colony-forming units/ml of the original water sample that were capable of growing on nutrient agar?

2. Three grams of soil were added to 27 ml of sterile water and shaken vigorously. After the soil particles settled, 0.1 ml of this was added to 9.9 ml of sterile water. This was further diluted by 4 successive 1/10

dilutions. One ml from the last dilution was used to prepare a pour plate. After incubation, 289 colonies were present on this plate. What was the number of colony-forming units/gram of the soil?

3. A bacterial culture was diluted and results from duplicate plates were obtained as indicated below. What was the number of colony-forming units/ml of the original culture?

Dilution used for plating	Amount plated	Colony counts after incubation (results from duplicate plates)
10^{-2}	0.1 ml	too many to count
10^{-3}	0.1 ml	too many to count
10^{-4}	0.1 ml	321; 403
10^{-5}	0.1 ml	34; 42
10^{-6}	0.1 ml	6; 1
10^{-7}	0.1 ml	0; 0
10^{-8}	0.1 ml	0; 0

4. Ten grams of hamburger were added to 90 ml of sterile buffer. This was mixed well in a blender. One-tenth of a ml of this slurry was added to 9.9 ml of sterile buffer. After thorough mixing, this suspension was further diluted by successive 1/100 and 1/10 dilutions. One-tenth of a ml of this final dilution was plated onto Plate Count Agar. After incubation, 52 colonies were present. How many colony-forming units were present in the *total* 10 gram sample of hamburger?

5. Devise a scheme to prepare a 10^{-6} dilution on a plate *using the least number of sterile water dilution tubes.*

6. Devise a scheme to prepare 1/20, 1/40 and 1/80 dilutions of a disinfectant.

Answers

1. $67 \times 10^4 = 6.7 \times 10^5$ (Remember to use plates with colony counts in the range of 30–300 colonies.)

2. $289 \times 10^7 = 2.9 \times 10^9$

3. $38 \times 10^6 = 3.8 \times 10^7$ (Remember to average duplicate plates.)

4. 5.2×10^9 (Did you remember to multiply by the total sample size of 10 grams?)

5. One possibility is to add 0.1 ml of sample to 99.9 ml of sterile water (that's a 1/1000), then 0.1 ml of this to 9.9 ml of sterile water (that's a 1/100, giving a total dilution of 10^{-5} at this point). Plate 0.1 ml of the second dilution to give 10^{-6} on the plate. The two first dilutions can be interchanged; the result will be the same. This scheme is not the only one possible with only two dilution tubes or bottles, however no scheme should use dilutions larger than 100 ml.

6. One possibility is to add 5 ml of the disinfectant to 95 ml of sterile water. This results in 5ml/(5ml + 95ml) = 5/100 = 1/20. Then dilute this by two successive 1/2 dilutions by mixing equal amounts of sterile water and previous dilution (for example, 5 ml of diluent + 5 ml of the previous dilution).

*For this reason, the results of cell enumeration in the viable plate count are given as colony-forming units/ml, not as cells/ml.

†The pour plate method of enumeration, where a sample is mixed with the molten agar and then the mixture is allowed to solidify in a plate, is generally used for sample sizes with volumes between about 0.5 to 1.5 ml. The spread plate technique, where a sample is spread on the surface of solidified agar, requires small sample sizes (for example, 0.1 ml) to allow the liquid to soak into the agar so that discrete colonies can be produced on the surface.

‡It is assumed that each student is familiar with scientific notation and the use of negative exponents. If you are not, consult your instructor.

Appendix **1**

Media, Reagent and Equipment Preparation

Media preparation

This appendix is designed to give general information about media preparation. The media formulations are given in Student Supplement 2, including special notes for preparation.

- Unless otherwise noted, use distilled water.
- Add ingredients to the water *in the order listed*.
- Be sure each chemical is dissolved before adding the next. A magnetic stirrer is useful for this. The teflon-coated stirring bar can be removed before sterilization or can be left in to facilitate stirring afterward.
- When weighing out dyes, be careful to avoid spills. If the medium will be made often, it is useful to prepare an aqueous or alcoholic stock solution of the dye, so that weighing of the dye crystals can be restricted to a few times a year.
- If the medium contains agar and must be dispensed before sterilizing, heat the medium in a 100° steamer or, *with constant stirring*, heat the medium to boiling on a hot plate. It generally requires about 20 minutes to completely dissolve 15 grams/liter of agar in a one liter quantity. Allow more time for larger volumes. Do not heat the medium longer than necessary. Mix the melted agar medium well before dispensing.
- If the medium contains agar and does not need to be dispensed before sterilizing, it can be placed directly in an autoclave. For volumes up to 3 liters, the agar will dissolve during the sterilizing process. (Larger volumes must be steamed to dissolve the agar before sterilizing.) After sterilizing, it is important to swirl the flask containing the medium to mix the layer of dissolved agar on the bottom of the flask. Try to avoid swirling so vigorously that bubbles are formed, since they will be transferred to the plates.
- Dispensing media into tubes is easier if an automatic pipetter or pipetting machine is used. (However, liquids being dispensed into tubes may splash, so the operator should be careful when dispensing hot liquids.) Some media require a specific volume of liquid in the tube for best results (see Student Supplement 2).
- With any medium which requires aseptic additions after autoclaving, the container used should be large enough to accommodate the additions and subsequent mixing.
- Any media to be sterilized by autoclaving should be placed in heat-re-

Safety Caution!

221

sistant Pyrex or Kimax containers.

- Since liquid expands when heated, the volume of liquid in any container to be sterilized in an autoclave should never exceed 80% of the container volume (for example, do not put more than 800 ml of liquid in a 1 liter flask).
- Tubes, bottles and flasks should be covered with *self-venting* closures, which allow for release of pressure during sterilization in an autoclave. Examples of self-venting closures are cotton or styrofoam plugs, plastic caps, screw caps left slightly loose, and foil.
- After sterilization, media which require further additions should be completed using aseptic procedures.

Preparation of glassware for sterilization

Glass petri plates and pipettes are best sterilized with dry heat in an oven for 4 hours at 425° F or overnight at 250° F. Because of the plastic or cotton closures used, glass tubes are best sterilized in the autoclave for 15 minutes at 121° C (250° F). The autoclave can be exhausted rapidly ("fast exhaust") when the tubes are empty.

Sterilization using the autoclave

An autoclave should be used whenever the medium or equipment to be sterilized is *not* heat-labile. Heat-labile ingredients are identified in the formulations given in Student Supplement 2. Heat-labile equipment includes most plastics and low-quality glass.

The autoclave sterilizes by raising the temperature inside the autoclave to 121° C using steam under pressure (about 15 psi above atmospheric pressure). The moist heat denatures proteins and other cellular constituents resulting in cell death. Since proteins are affected, viruses are also destroyed in this process. To ensure complete sterility, the material must be kept at 121° C for 15 minutes. Since the thermometer of the autoclave measures the temperature in the *air-space* in the autoclave, the required time in the autoclave may exceed 15 minutes after the thermometer has reached 121°, to compensate for the time it takes for the *material* inside to reach 121°. Typically, 15–20 minutes of autoclaving is sufficient for small loads in an autoclave, and more time is needed if either the total amount of material in the autoclave or the volume of any container (for example, a 6-liter container) is large.

Steps for autoclaving are outlined below:

- Place material to be sterilized in the autoclave. Be sure all containers are heat- and pressure-resistant and that all closures are self-venting. Containers with liquids should not be more than 80% full. To allow maximum heat exchange, allow enough space between each container for steam to contact all sides.
- Close the autoclave door firmly. If the door is closed properly, no steam or liquid should escape during the actual autoclaving.
- Read and follow the directions carefully for use of your particular autoclave. If the autoclave has automatic controls, set the timer for the length of time appropriate for the volume of material to be autoclaved (but *not*

for less than 15 minutes) and set the exhaust routine for "slow exhaust" for loads containing liquids or "fast exhaust" for dry loads such as empty tubes. Slow exhaust is required for liquids because during sterilization the liquid is at 121°, (and thus 21° above its boiling point at atmospheric pressure). Slow exhaust allows the liquid to cool below its boiling point as the pressure descends. Using fast exhaust with liquids will result in loss of the liquid from the container as the pressure descends and the super-heated liquid boils over.

- Turn on the steam (or if automatic, set to start). Begin timing when the temperature reaches 121°.
- For non-automatic autoclaves: after the appropriate period of time, turn off the steam. For slow exhaust, do not open the exhaust vent; for fast exhaust, open it.
- Allow the autoclave to exhaust properly (indicated by a return to 0 on the pressure gauge on non-automatic autoclaves or by a signal on automatic ones). Open the autoclave *carefully*, keeping the arms and face away from the opening to avoid being burned by any steam that may leave the autoclave as the door is opened.

Safety Caution!

- *Using properly insulated gloves*, remove the material from the autoclave. *Do not* remove any containers while the liquid is still boiling or bumping, as the liquid in these containers may boil out of the container and cause a burn. Place materials in an appropriate area for cooling. Any materials which do not need to be treated further prior to use should be placed in an area away from normal use to avoid possible burns to personnel. Tubes which contain media for agar slants should be tilted in baskets or racks for cooling. Flasks containing media for plates should be placed in a 50–55° water bath until cool enough for pouring.

Safety Caution!

Sterilization by filtration

Filtration is used to sterilize heat-labile liquids such as antibiotics or vitamins. The most commonly used pore size for sterilization of liquids for media is 0.45 μm, although 0.2 μm pores may be used in special situations. It is important to note that filter sterilization *does not remove viruses*. Viruses, however, are rarely a problem in media preparation.

The filter, filter holder, and receiving container must all be sterile. Presterilized, disposable filter units are available from a large number of vendors of microbiological equipment. Re-usable filter holders and autoclavable filters are also available. Both types of filter holders are available for use either with a syringe or with a vacuum source.

Reagent preparation

- Materials to be mixed for reagents should be added to the solvent (for example, water, alcohol, acetone) *in the order listed*.
- Weigh dyes carefully to avoid spilling the crystals in the weighing area. Any spilled dye crystals may subsequently contaminate other media and reagents.
- Prepare all reagents containing volatile liquids *in a properly functioning chemical hood*. Volatile liquids include, but are not limited to, acetone,

Safety Caution!

butanol, concentrated HCl or other concentrated acids, and toluene. Eye protection should be worn.

Safety
Caution!

- Mix reagents containing concentrated acids or bases carefully. Be aware that many of these mixtures will become hot, so proper heat-resistant containers should be used, and gloves should be available. Always add the acid or the base *to the solvent*, not the other way around, to prevent possible explosion.
- Many reagents are light sensitive, so stock solutions of reagents should routinely be stored in dark or opaque bottles.
- Clean up all reagent spills immediately.

Culture Preparation

The cultures used in this manual are generally easy to handle. In most cases, the exercise requires a liquid culture of easily grown bacteria. However, a few exercises require more elaborate culture preparation. Table A2-1 is a list of the cultures used, the media they usually should be grown in, the temperature and time of incubation, and the exercises in which they are used. Included in Table A2-1 are the American Type Culture Collection (ATCC) catalog* numbers for ordering the cultures. These numbers are included for ordering convenience, but are not meant to imply that other cultures of the same organism which might already be available are not usable. If in doubt as to whether an available culture will work, refer to the particular exercise in which the organism is used and perform the required tests on the culture.

Table A2-1   Cultures used in this manual

Cultures	ATCC number[1]	Exercises used in	Growth medium[2]	Growth temperature[3]	Special notes
Bacteria					
Bacillus cereus	e14579	1,3	Nutrient broth	30–37°	
Bacillus megaterium	e14581	2	Nutrient broth	30–37°	
Bacillus polymyxa	10401	11,17	Nutrient broth	30–37°	
Bacillus subtilis	e6051	6,11,14,23	Nutrient broth	30–37°	
Branhamella catarrhalis	25238	21	Brain heart infusion broth	37°	
Clostridium butyricum	e8260	6	Fluid thioglycollate broth	30–37°	Strict anaerobe
Enterobacter aerogenes	e13048	22	Nutrient broth	30–37°	
Escherichia coli B	e11303	14	Nutrient broth	30–37°	
Escherichia coli F⁻	e25252	(9),12	Nutrient broth	37°	
Escherichia coli Hfr	e23739	12	Nutrient broth	37°	
Escherichia coli K	e23716	many	Nutrient broth	30–37°	Can substitute *E. coli B*, except in Exercise 14
Klebsiella pneumoniae	15574	3,8,10,11,14,23	Nutrient broth	30–37°	Grow on EMB agar for Exercise 3
Lactobacillus bulgaricus	11842	6,9,11,19,23	All purpose Tween broth	37°	Grow in sterile skim milk for Exercise 19
Leuconostoc mesenteroides	27258	6	All purpose Tween broth	37°	
Micrococcus luteus	e4698	8,11,23	Nutrient broth	30–37°	

Table A2-1 continued

Cultures	ATCC number[1]	Exercises used in	Growth medium[2]	Growth temperature[3]	Special notes
Mycobacterium smegmatis	e14468	(3)	Brain heart infusion broth	37°	Grows slowly, allow 2–3 days
Neisseria sicca	29256	21,23	Brain heart infusion broth	37°	
Proteus vulgaris	e13315	3,22,23	Nutrient broth	30–37°	
Pseudomonas fluorescens	e13525	6,13,23	Nutrient broth	30°	
Spirillum volutans	19554	3	Peptone-succinate broth	30°	
Staphylococcus aureus	e12600	21	Nutrient broth	30–37°	
Staphylococcus epidermidis	14990	1,3,7,8,10,11,12,21,23	Nutrient broth	30–37°	
Streptococcus faecalis	19433	21,23	All purpose Tween broth	37°	
Streptococcus faecalis subsp. *zymogenes*	6055	21	All purpose Tween broth	37°	
Streptococcus mitis	903	21	All purpose Tween broth	37°	
Streptococcus salivarius	25975	21	All purpose Tween broth	37°	
Streptococcus thermophilus	19258	19	Skim milk	37°	
Streptomyces griseus	e10137	6,17	Streptomyces agar	30°	Does not grow well in liquid culture. Grow in solid culture; suspend spores in sterile 10^{-2} M potassium phosphate buffer, pH 7.0, containing 0.001% sodium lauryl sulfate

Fungi

Cultures	ATCC number[1]	Exercises used in	Growth medium[2]	Growth temperature[3]	Special notes
Aspergillus carbonarius	1025	4	Sabouraud dextrose agar	25–30°	Grows slowly, allow 5–7 days
Penicillium chrysogenum	9480	4	Sabouraud dextrose agar	25–30°	Grows slowly, allow 5–7 days
Rhizopus nigricans	24795	4	Sabouraud dextrose agar	25–30°	Grows slowly, allow 5–7 days
Saccharomyces cerevisiae	287	2,4,16	Nutrient agar or broth containing 1% glucose	25–30°	

Viruses

Cultures	ATCC number[1]	Exercises used in	Growth medium[2]	Growth temperature[3]	Special notes
Bacteriophage T4	11303-B4	14	*E. coli* B		See preparation of bacteriophage stocks in this appendix
Bacteriophage T2	e11303-B2	14	*E. coli* B		

[1]Strain numbers from the ATCC Catalog of Strains 1, 15th edition, 1982. Catalog is available free from: American Type Culture Collection, 12301 Parklawn Drive, Rockville, Maryland, 20852–1776.

[2]Unless otherwise specified in the exercise.

[3]Unless otherwise specified in the exercise. If a range of temperature is given, incubate at the lower temperature for slower growth.

Maintenance of stock cultures

Because of the expense and trouble of obtaining microbial cultures, it is useful to maintain stock cultures. For short-term maintenance, an agar slant culture can be prepared and refrigerated (for up to 3 months, depending on the organism, see Table A2-2). Long-term maintenance can be accomplished by freezing (at $-70°$ or in liquid nitrogen) or freeze-drying (lyophilizing) the culture. Since freeze-drying requires specialized equipment, it will not be covered here. In some cases, cultures can be stored frozen in a conventional freezer at $-20°$.

To prepare a culture for freezing, a few different techniques can be used.

Table A2-2 Culture maintenance

Culture	Medium[1]	Incubation period before placing into storage	Length of time before transfer is necessary[2]
All species of Enterobacter Escherichia Klebsiella Micrococcus Proteus Pseudomonas Staphylococcus	Nutrient agar + 1% glycerol slants	1–2 days	3–6 months at 4°; 1–2 years at $-20°$[3]
Bacillus Streptomyces	Nutrient agar slant	3–5 days	1–2 years at 25°; 3–4 years at $-20°$
Branhamella Mycobacterium Neisseria	Brain heart infusion agar slant	1–2 days	3–4 months at 4°; 1 year at $-20°$
Clostridium	Fluid thioglycollate broth + 0.5% $CaCO_3$[4]	3–5 days	1–2 years at 4°; 3–4 years at $-20°$
All species of Lactobacillus Leuconostoc Streptococcus	All purpose Tween slant + 0.5% $CaCO_3$,[4] with mineral oil added after growth	1–2 days	3–4 months at 4°; 1 year at $-20°$
Spirillum	Peptone-succinate agar slant	2–3 days	1 month at 25°; 1 year at $-20°$
Aspergillus Penicillium Rhizopus	Sabouraud dextrose agar slant	5–7 days, until spores are seen	1 year at 4°; 3–4 years at $-20°$
Saccharomyces	Sabouraud dextrose agar + 0.5% $CaCO_3$[4]	1–2 days	1 year at 4°; 3–4 years at $-20°$

[1]All media should be made in screw-capped tubes which will seal against moisture loss over the storage period.

[2]It is often necessary to transfer the culture several times to obtain peak viability before returning the culture to storage conditions.

[3]A temperature of $-80°$ is preferred. Many freezers that are part of a refrigerator will not maintain temperatures as low as $-20°$. Check the specific freezer before using.

[4]$CaCO_3$ is added as a solid to neutralize acids formed during the growth of the culture. The $CaCO_3$ will be a white precipitate in uninoculated media; mix in well during distribution of the media into tubes.

1. The culture can be grown on a slant of nutrient agar to which 1% glycerol has been added. The slant should be in a screw-capped tube. After incubation, tighten the cap and place the tube in a freezer until needed. The glycerol in the medium may protect the cells from injury during freezing. It is important that the cap be sealed well or desiccation will occur. To use such a culture, scrape the surface of the still-frozen or thawed slant† with a loop and inoculate into a rich liquid medium. In most cases, the culture will grow well within 24 hours of incubation, as indicated by turbidity of the broth.

2. The culture can be grown up in a liquid medium to which sterile glycerol is added to a final concentration of 10% before freezing. The culture can then be dispensed in small amounts in screw-capped tubes and placed into a freezer until needed. For use, thaw one tube of culture and pipette the contents into a ten-fold larger volume of a rich liquid medium. Recovery should be rapid, and turbidity can usually be seen in 6–12 hours.

3. The culture can be grown on an agar slant which can be covered with sterile mineral oil before freezing. After freezing, the mineral oil is removed by a pipette and cells scraped from the slant and treated as in #1 above. This is not a recommended procedure for strict aerobes.

Preparation of liquid cultures

Liquid cultures are easily prepared in large volume, then dispensed in small volumes for students just before use. Inoculate a flask of sterile medium (100–150 ml in a 500 ml flask) with 1–2 ml of culture from a liquid stock or with a loopful of cells from a culture grown on solid medium. For aerobes, incubate overnight on a shaker at the appropriate temperature, adjusting the shaker speed to obtain maximum aeration without splashing. For aerotolerant anaerobes, such as the lactic acid bacteria, shaking is not necessary. Dispense the grown culture into the individual tubes for distribution with a sterile pipette or automatic manual pipetter.

Preparation of hay infusions

Hay infusions can be prepared by adding hay to water (1/3 volume hay, 2/3 volume lake, pond, or river water), then incubating this mixture in the light for 5–7 days. The hay should be fresh and cut into small (5–10 cm) pieces. Add distilled water periodically to compensate for evaporation.

Preparation of Henrici slides

Henrici slides can be prepared for microscopic examination of the fungi using the medium indicated in Table A2-1. Sterile slides and coverslips are required. These can be sterilized either by autoclaving inside glass petri dishes or by flame-sterilization as done with glass spreaders. *Caution!* Coverslips shatter easily when sterilized by flame. Wear eye protection!

To make the Henrici slide:

1. Place a drop of melted agar medium in the center of a sterile slide and allow it to solidify.
2. Sterilize your loop, and then touch the *hot* loop to sterile agar to cool.

This will result in a fine film of agar on the loop to which fungal spores will adhere. Obtain the inoculum by drawing this loop across the fungal culture.

3. Using the inoculated loop, cut the agar drop on the slide in two, pushing the two pieces about 2–3 mm apart. The edges of the cut agar are now inoculated.

4. Place the sterile coverslip on the agar drop. Press lightly to assure good contact between the coverslip and the agar. The culture to be observed will grow out laterally at the cut edges of the agar drop.

5. The Henrici slide is incubated in a moist chamber. To prepare the moist chamber, aseptically place two sheets of sterile filter paper in the bottom of a petri dish and add just enough sterile water to moisten the paper. Place the Henrici slide in the moist chamber, seal the edges of the plate to prevent drying, and incubate at 30° for 5–7 days.

6. Observe growth and sporulation in the space between the two halves of the cut agar. The coverslip is *not* removed for this. Since a Henrici slide is thicker than a normal slide, students should use caution when using the high-dry or oil-immersion lenses, to avoid damage to the lens and to the slide preparation.

Preparation of cultures of *Clostridium*, an anaerobe

The easiest way to obtain anaerobic conditions for clostridia is to prepare the culture in fluid thioglycollate broth. This broth is available commercially, and has the same formulation as the thioglycollate agar used in Exercise 6, except that the agar concentration has been reduced to 0.75 g/l. Because this medium is difficult to dispense when cool, it is recommended that individual cultures be grown in tubes containing 6–8 ml of medium.

Preparation of concentrated water samples

Water samples can easily be concentrated using membrane filtration. For example, 100 ml of water can be added to the top of a large-volume filtration system and vacuum applied. If the water sample is to be concentrated 10-fold, the vacuum is broken when the volume of liquid left in the top is down to 10 ml. The liquid is then swirled to dislodge large organisms from the surface of the filter, and the sample removed.

Preparation of bacteriophage stocks

Bacteriophage stocks can be prepared in a variety of ways. The method requiring the least amount of equipment involves inoculating a tube of top agar with host cells and a large number of virus particles to give a multiplicity of infection of 1:1,000–10,000 (one virus particle for every 1,000–10,000 host cells). An overnight culture of *E. coli* contains approximately 10^9 bacteria/ml. Ascertain the virus titer beforehand by plaque assay. Pour the top agar onto a plate of bottom agar, and incubate at the appropriate temperature (37° for *E. coli*) overnight.

Following incubation, the plate should be clear, if the number of virus particles was high enough. Add 5 ml of sterile 10^{-2} M potassium phosphate buffer, pH 7, to the plate. Using a sterile spreader, lightly agitate the surface of the agar. Using a sterile pasteur pipette, remove the liquid. It should contain 10^5 to 10^8 viruses per ml. To ensure the absence of viable host cells, either pass this liquid through a sterile membrane filter or add one drop of chloroform to the virus suspension and let it sit in the refrigerator for 2 days. *Always* check for the absence of host cells in this stock by plating 0.1 ml of it onto nutrient agar.

An alternative method is to inoculate the virus and host into liquid culture, incubate for 12–24 hours, then remove any remaining host cells by filtration or, if the cells are numerous, by centrifugation followed by filtration. This method is often more efficient, but usually requires a centrifugation step.

Preparation of unknown cultures

Unknowns containing only one pure culture are prepared like a normal culture. Mixed unknowns, however, are treated differently. After growth in a liquid medium, each pure culture is diluted 1/10 in sterile saline (0.85% NaCl). This effectively dilutes out any endproducts from the previous growth cycle which might be toxic to the organism(s) with which the culture is to be mixed. Next combine the cultures together in a separate sterile flask. The ratio of Gram-positive to Gram-negative organisms should be 10/1 to 15/1, or the Gram-positive organism will be difficult to recover. Mixtures should not be kept for more than 4–6 hours before use, as some of the organisms will begin to die due to the presence of toxic products from other organisms.

*The American Type Culture Collection catalog is available from ATCC at 12301 Parklawn Drive, Rockville, Maryland, 20852-1776. The catalog is shipped free at book rate. This organization sells at reduced rate a range of cultures suitable for teaching. The code numbers of these cultures are preceded by the letter "e."

†Note that the agar will appear "shattered" as the agar polymer separates from the water of the medium during freezing and thawing.

Time Required for Each Exercise

The following table lists the *approximate* times required for a student to complete the periods in each exercise. This table is intended to assist the instructor with developing a laboratory schedule.

Exercise	Period	Approximate Completion Time
1	A	45 min–1 hr
	B	30–45 min
	C	10 min
2	A	1–1½ hr
	B	45 min–1 hr
3	A	1½ hr
	B	1½ hr
	C	1 hr
4	A	2 hrs
5	A	1½ hr
	B	1 hr
6	A	1 hr
	B	2 hr
7	A	30–45 min
	B	30–45 min
8	A	45 min–1 hr
	B	45 min
	C	1 1½ hr
9	A	20 min
	B	45 min–1 hr
10	A	20 min
	B	1–1½ hr
11	A	45 min–1 hr
	B	1–1½ hr
12	A	1–1½ hr
	B	45 min–1 hr
	C	45 min
	D	20 min
13	A	15 min
	B	30–45 min
14	A	45 min
	B	30 min
15	A	1 hr
16	A	10 min
	B	20–30 min
	C	30 min
17	A	20 min
	B	30 min
	C	1½ hr
	D	20 min
	E	10 min
	F	15 min

(Continued)

Exercise	Period	Approximate Completion Time
18	A	45 min–1 hr
	B	30 min
19	A	1½–2 hr
	B	1–1½ hr
	C	30 min
20	A	30–45 min
	B	30–45 min
21	A	1½–2 hrs
	B	1½–2 hrs
	C	30–45 min
22	A	1–1½ hr
	B	1 hr
	C	30 min
	D	1 hr
	E	20–30 min
23	A	none in class
	B	30 min
	C	30–45 min
	D	1 hr
	E	1 hr

List of Vendors

This list of vendors is supplied to assist in the acquisition of equipment and supplies for the exercises in this manual. This list is for convenience only and neither implies endorsement of nor intends discrimination against any vendors.

American Scientific Products
  1210 Waukegan Road
  McGaw Park, IL 60085
  800-323-4515

Equipment such as water baths and pipette bulbs, as well as sterile disposable pipettes and filtration devices

American Type Culture Collection
  12301 Parklawn Drive
  Rockville, MD 20852
  800-638-6597

Microbial cultures

Bellco
  P. O. Box B
  340 Edrudo Road
  Vineland, NJ 08360
  800-257-7043

Biological equipment and glassware

BioQuest (BBL)
  P. O. Box 243
  Cockeysville, MD 21030
  301-666-0100

Dehydrated media and other microbiological supplies

Calbiochem-Behring Corporation
  P. O. Box 12087
  San Diego, CA 92112
  800-854-2171

Special chemicals such as ortho-nitrophenyl-β,D-galactoside

Difco Laboratories
  P. O. Box 1058A
  Detroit, MI 48232
  313-961-0800

Dehydrated media, antisera, and other microbiological supplies

Fisher Scientific
  711 Forbes Avenue
  Pittsburgh, PA 15219
  412-562-8300

Equipment, chemicals and microbiological supplies

Gibco Diagnostics
  P. O. Box 4385
  Madison, WI 53711
  608-221-2221

Prepared media and other microbiological supplies

Polysciences, Inc.
  400 Valley Road
  Warrington, PA 18976
  800-523-2575

Equipment and chemicals

(Continued on next page)

Sigma Chemical Company
  P. O. Box 14508
  St. Louis, MO 63178
  800-325-3010

VWR Scientific, Inc.
  P. O. Box 3200
  San Francisco, CA 94119
  415-468-7150

Special chemicals such as ortho-nitrophenyl-β,D-galactoside

Equipment, chemicals and microbiological supplies

Notes

Notes

Notes

Notes

Notes

Notes

Notes

Notes

Notes

Notes